Elementary FORGE PRACTICE

JOHN LORD BACON

Lost Technology Series

Reprinted by Lindsay Publications Inc.

Elementary

Forge-Practice

John Lord Bacon

Second edition, enlarged, published by John Wiley & Sons, Inc., New York, 1914.

ISBN 0-917914-45-7

2 3 4 5 6 7 8 9 0

FORGE-PRACTICE

(*ELEMENTARY*)

BY

JOHN LORD BACON

Construction Engineer

Member American Society of Mechanical Eugineers

Sometimé Instructor in Forge Practice and Machine Design, Lewis Institute, Chicago

SECOND EDITION, ENLARGED

TOTAL ISSUE, NINE THOUSAND

WARNING

Remember that the materials and methods described here are from another era. Workers were less safety conscious then, and some methods may be downright dangerous. Be careful! Use good solid judgement in your work, and think ahead. Lindsay Publications, Inc. has not tested these methods and materials and does not endorse them. Our job is merely to pass along to you information from another era. Safety is your responsibility.

Write for a catalog or other unusual books available from:

Lindsay Publications, Inc.
PO Box 12
Bradley, IL 60915-0012

To my Wife,

WITHOUT WHOSE ASSISTANCE IT WOULD
NEVER HAVE BEEN WRITTEN,
THIS LITTLE VOLUME IS
DEDICATED.

PREFACE.

This little volume is the outgrowth of a series of notes given to the students at Lewis Institute from time to time in connection with shop work of the character described.

It is not the author's purpose to attempt to put forth anything which will in any way take the place of actual shop work, but rather to give some explanation which will aid in the production of work in an intelligent manner.

The examples cited are not necessarily given in the order in which they could most advantageously be made as a series of exercises, but are grouped under general headings in such a way as to be more convenient for reference.

The original drawings from which the engravings were made were drawn by L. S. B.

PREFACE TO THE SECOND EDITION.

THE author believes that the text book should be used to explain principles and give examples, not to give minute explicit directions for making a set of exercises.

This necesitates an independent set of drawings for the work to be done.

It is for this purpose that the set of drawings is given.

The author has felt the need and lack of such drawings in the text-book as used before, and it is to remedy this defect that the addition has been made.

The exercises are such as have been found useful in the shop, and an effort has also been made to give drawings of such tools as are ordinarily used in the forge and machine shops.

J. L. B.

March, 1908.

CONTENTS.

CHAPTER IX.

CHAPTER X.

CHAPTER XI.

CHAPTER XII.

FORGE-PRACTICE.

CHAPTER I.

GENERAL DESCRIPTION OF FORGE AND TOOLS.

Forge.—The principal part of the forge as generally made now is simply a cast-iron hearth with a bowl, or depression, in the center for the fire. In the bottom of this bowl is an opening through which the blast is forced. This blast-opening is known as the tuyere. Tuyeres are made in various shapes; but the object is the same in all, that is, to provide an opening, or a number of openings, of such a shape as to easily allow the blast to pass through, and at the same time, as much as possible, to prevent the cinders from dropping into the blast-pipe.

There should be some means of opening the blast-pipe beneath the tuyere and cleaning out the cinders which work through the tuyere-openings, as some cinders are bound to do this no matter how carefully the tuyere is designed.

When a long fire is wanted, sometimes several

tuyeres are placed in a line; and for some special work the tuyeres take the form of nozzles projecting inwardly from the side of the forge.

Coal.—The coal used for forge-work should be of the best quality bituminous, or soft, coal. It should coke easily; that is, when dampened and put on the fire it should cake up, form coke, and not break into small pieces. It should be as free from sulphur as possible, and make very little clinker when burned.

Good forge-coal should be of even structure through the lumps, and the lumps should crumble easily in the hand. The lumps should crumble rather than split up into layers, and the broken pieces should look bright and glossy on all faces, almost like black glass, and show no dull-looking streaks.

Ordinary soft coal, such as is used for "steaming-coal," makes a dirty fire with much clinker. "Steaming-coal" when broken is liable to split into layers, some of which are bright and glossy, while others are dull and slaty-looking.

Fire.—On the fire, to a very great extent, depends the success or failure of all forging operations, particularly work with tool-steel and welding.

In building a new fire the ashes, cinders, etc., should be cleaned away from the center of the forge down to the tuyere. Do not clean out the whole top of the forge, but only the part where the new fire is wanted, leaving, after the old material has been taken out, a clean hole in which to start the fresh fire.

The hearth of the forge is generally kept filled with cinders, etc., even with the top of the rim.

Shavings, oily waste, or some other easily lighted material should be placed on top of the tuyere and set on fire.

As soon as the shavings are well lighted, the blast should be turned on and coke (more or less of which is always left over from the last fire) put on top and outside of the burning shavings. Over this the "green coal" should be spread.

Green coal is fresh coal dampened with water. Before using the forge-coal it should be broken into small pieces and thoroughly wet with water. This is necessary, as it holds together better when coking, making better coke and keeping in the heat of the fire better. It is also easier to prevent the fire from spreading out too much, as this dampened coal can be packed down hard around the edges, keeping the blast from blowing through.

The fire should not be used until all the coal on top has been coked. As the fire burns out in the center, the coke, which has been forming around the edge, is pushed into the middle, and more green coal added around the outside.

We might say the fire is made up of three parts: the center where the coke is forming and the iron heating; a ring around and next to this center where coke is forming; and, outside of this, a ring of green coal.

This is the ordinary method of making a small fire.

This sort of fire is suitable for smaller kinds of

work. It can be used for about an hour or two, at the end of which time it should be cleaned. When welding, the cleaning should be done much oftener.

Large Fires.—Larger fires are sometimes made as follows: Enough coke is first made to last for several hours by mounding up green coal over the newly started fire and letting it burn slowly to coke thoroughly. This coke is then shoveled to one side and the fire again started in the following way: A large block, the size of the intended fire, is placed on top of the tuyere and green coal is packed down hard on each side, forming two mounds of closely packed coal. The block is taken out and the fire started in the hole between the two mounds, coke being added as necessary. This sort of a fire is sometimes called a stock fire, and will last for some time. The mounds keep the fire together and help to hold in the heat.

For larger work, or where a great many pieces are to be heated at once, or when a very even or long-continued heat is wanted, a furnace is used. For furnace use, and often for large forge-fires, the coke is bought ready-made.

Banking Fires. — When a forge-fire is left it should always be banked. The coke should be well raked up together into a mound and then covered with green coal. This will keep the fire alive for some time and insure plenty of good coke for starting anew when it does die out. A still better method to follow, when it is desired to keep the fire for some time, is to bury a block of wood in the center of the fire when banking it.

Oxidizing Fire.—When the blast is supplied from a power fan, or blower, the beginner generally tries to use too much air and blow the fire too hard.

Coal requires a certain amount of air to burn properly, and as it burns it consumes the oxygen from the air. When too much blast is used the oxygen is not all burned out of the air and will affect the heated iron in the fire. Whenever a piece of hot iron comes in contact with the air the oxygen of the air attacks the iron and forms oxide. This oxide is the scale which is seen on the outside of iron. The higher the temperature to which the iron is heated, the more easily the oxide is formed. When welding, particularly, there should be as little scale, or oxide, as possible, and to prevent its formation the iron should not be heated in contact with any more air than necessary. Even on an ordinary forging this scale is a disadvantage, to say the least, as it must be cleaned off, and even then is liable to leave the surface of the work pitted and rough. If it were possible to keep air away from the iron entirely, no trace of scale would be formed, even at a high heat.

If just enough air is blown into the fire to make it burn properly, all the oxygen will be burned out, and very little, if any, scale will be formed while heating. On the other hand, if too much air is used, the oxygen will not all be consumed and this unburned oxygen will attack the iron and form scale. This is known as "oxidizing";

that is, when too much air is admitted to the fire the surplus oxygen will attack the iron, forming "oxide," or scale. This sort of a fire is known as an "oxidizing" fire and has a tendency to "oxidize" anything heated in it.

Anvil.—The ordinary anvil, Fig. 1, has a body of cast iron, wrought iron, or soft steel, with a tool-steel face welded on and hardened. The hardened steel covers just the top face, leaving the horn and the small block next the horn of the softer material.

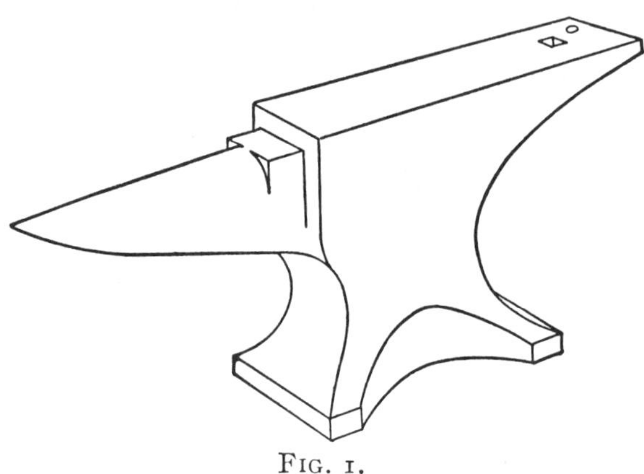

FIG. 1.

The anvil should be so placed that as the workman faces it the horn will point toward his left.

The square hardie-hole in the right-hand end of the face is to receive and hold the stems of hardies, swages, etc.

For small work the anvil should weigh about 150 lbs.

Hot and Cold Chisels.—Two kinds of chisels are commonly used in the forge-shop: one for cutting cold stock, and the other for cutting red-hot metal. These are called cold and hot chisels.

The cold chisel is generally made a little thicker in the blade than the hot chisel, which is forged down to a thin edge.

Fig. 2 shows common shapes for cold and hot chisels, as well as a hardie, another tool used for cutting.

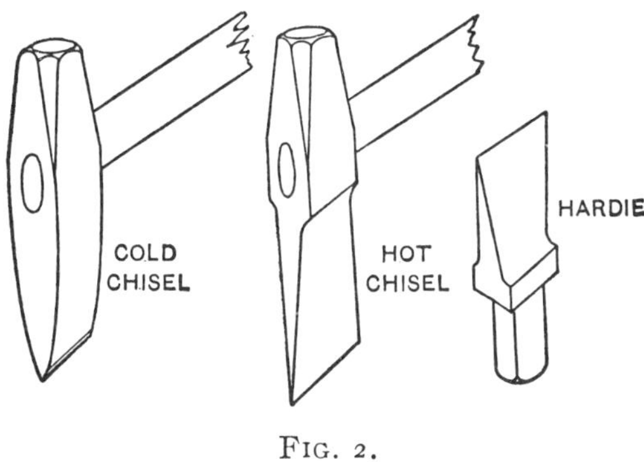

Fig. 2.

Both chisels should be tempered alike when made.

The cold chisel holds its temper; but, from contact with hot metal, the hot chisel soon has its edge softened. For these reasons the two chisels should never be used in place of each other, for by using the cold chisel on hot work the temper is drawn and the edge left too soft for cutting cold metal, while the hot chisel soon becomes so soft that if used in place of the cold it will have its edge turned and ruined.

It would seem that it is useless to temper a hot chisel as the heated work, with which the chisel comes in contact, so soon draws the temper. When the chisel is tempered, however, the steel is left in a much better condition even after being affected by hot metal on which it is used than it would be if the chisel were made untempered.

Grinding Chisels.—It is very important to have the chisels, particularly cold chisels, ground correctly, and the following directions should be carefully followed.

The sides of a cold chisel should be ground to form an angle of about 60° with each other, as shown in Fig. 3. This makes an angle blunt

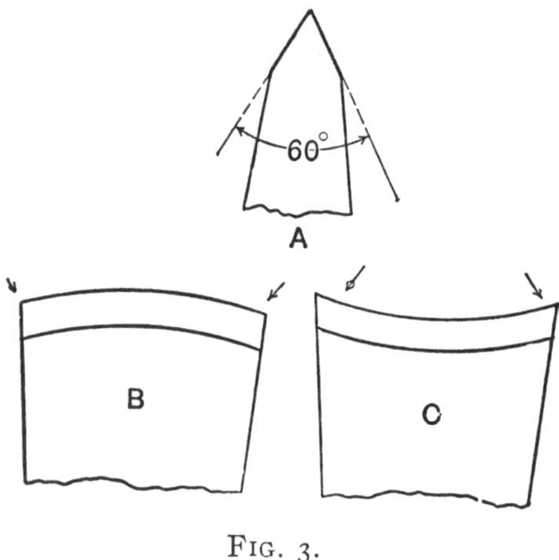

FIG. 3.

enough to wear well, and also sharp enough to cut well.

The cutting edge should be ground convex, or curving outward, as at *B*. This prevents the corners from breaking off. When the edge of the chisel is in this shape, the strain of cutting tends to force the corners back against the solid metal in the central part of the tool. If the edge were made concave, like *C*, the strain would tend to force the corners outward and snap them off. The arrows on *B* and *C* indicate the direction of these forces.

Hot chisels should be ground sharper. The sides should be ground at an angle of about 30° instead of 60°.

Another tool used for cutting is the hardie. This takes the place of the cold or hot chisel. It has a stem fitted to the square hole in the right-hand end of the anvil face, this stem holding the hardie in place when in use.

Cutting Stock.—When soft steel and wrought-iron bars are cut with a cold chisel the method should be about as follows: First cut about one-fourth of the way through the bar on one side; then make a cut across each edge at the ends of the first cut; turn the bar over and cut across the second side about one-fourth the way through; tilt the bar slightly, with the cut resting on the outside corner of the anvil, and by striking a sharp blow with the sledge on the projecting end, the piece can generally be easily broken off.

Chisels should always be kept carefully ground and sharp.

A much easier way of cutting stock is to use bar shears, but these are not always at hand.

The edge of a chisel should never under any circumstances be driven clear through the stock and allowed to come in contact with the hard face of the anvil.

Sometimes when trimming thin stock it is convenient to cut clear through the piece; in this case the cutting should be done either on the horn, the soft block next the horn, or the stock to be cut should be backed up with some soft metal. An easy way to do this is to cut a wide strip of stock about two inches longer than the width of the face of the anvil, and bend the ends down to fit over the sides of the anvil. The cutting may be done on this without injury to the edge of the chisel. It is very convenient to have one of these strips always at hand for use when trimming thin work with a hot chisel.

The author has seen a copper block used for this same purpose. The block was formed like Fig. 4, the stem being shaped to fit into the hardie-hole of the anvil. This block was designed for use principally when trimming thin parts of heated work with a hot chisel.

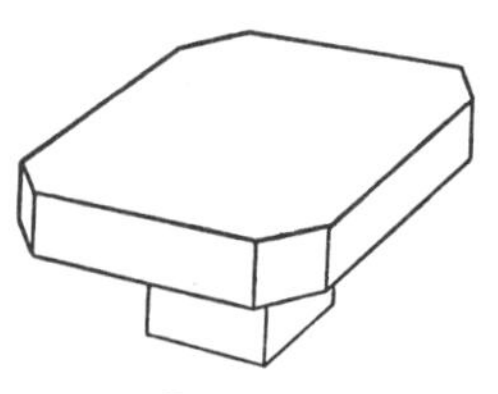

Fig. 4.

Care should always be taken to see that the work rests flat on the anvil or block when cutting. The work should be supported directly underneath the point where the cutting is to be done; and the solider the support, the easier the cutting.

Hammers.—Various shapes and sizes of hammers are used, but the commonest, and most convenient for ordinary use, is the ball pene-hammer shown in Fig. 5.

The large end is used for ordinary work, and the small ball end, or pene, for riveting, scarfing, etc. These hammers vary in weight from a few ounces up to several pounds. For ordinary use about a $1\frac{1}{2}$- or 2-pound hammer is used.

Several other types in ordinary use are illustrated

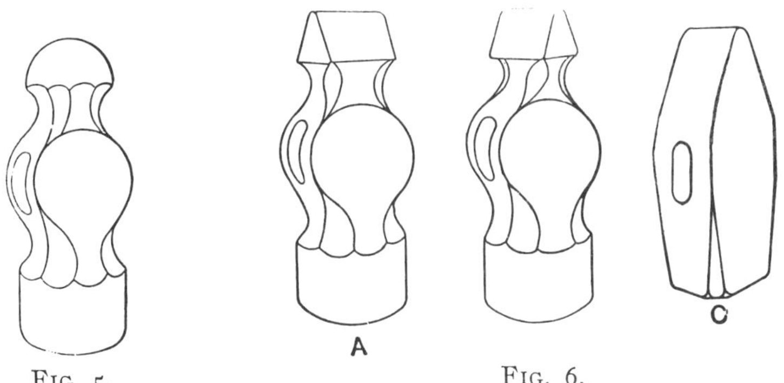

FIG. 5. FIG. 6.

in Fig. 6. *A* is a straight-pene; *B*, a cross-pene; and *C*, a riveting-hammer.

Sledges.—Very light sledges are sometimes made the same shape as ball-pene hammers. They are used for light tool-work and boiler-work.

Fig. 7 illustrates a common shape for sledges. This is a double-faced sledge.

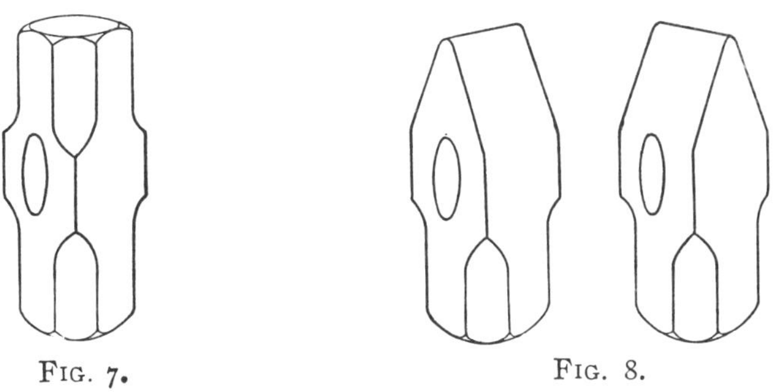

FIG. 7. FIG. 8.

Sledges are also made with a cross-pene or straight-pene, as shown in Fig. 8.

For ordinary work a sledge should weigh about 10 or 12 lbs.; for heavy work, from 16 to 20.

Sledges for light work weigh about 5 or 6 lbs.

Tongs.—Tongs are made in a wide variety of shapes and sizes, depending upon the work they are intended to hold. Three of the more ordinary shapes are illustrated.

The ordinary straight-jawed tongs are shown in Fig. 9. They are used for holding flat iron. For holding round iron the jaws are grooved or bent to the shape of the piece to be held.

Fig. 10 shows a pair of bolt-tongs. These tongs are used for holding bolts or pieces which are larger on the end than through the body, and are so shaped that the tongs do not touch the enlarged end when the jaws grip the body of the work.

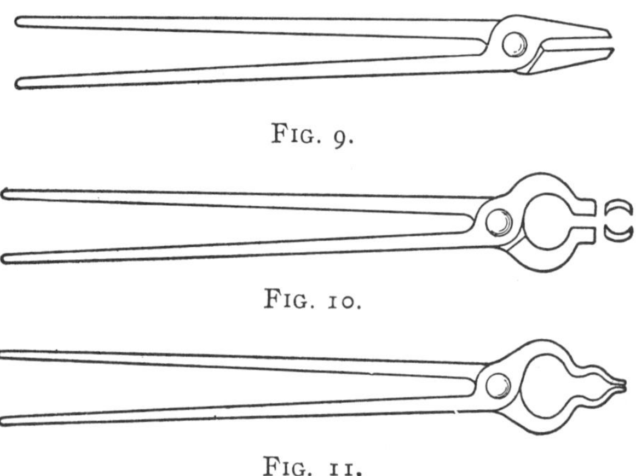

FIG. 9.

FIG. 10.

FIG. 11.

Pick-up tongs, Fig. 11, are used for handling small pieces, tempering, etc., but are very seldom used for holding work while forging.

Fitting Tongs to Work.—Tongs should always be carefully fitted to the work they are intended to hold.

Tongs which fit the work in the manner shown in Fig. 12 should not be used until more carefully fitted. In the first case shown, the jaws are too close together; and in the second case, too far apart.

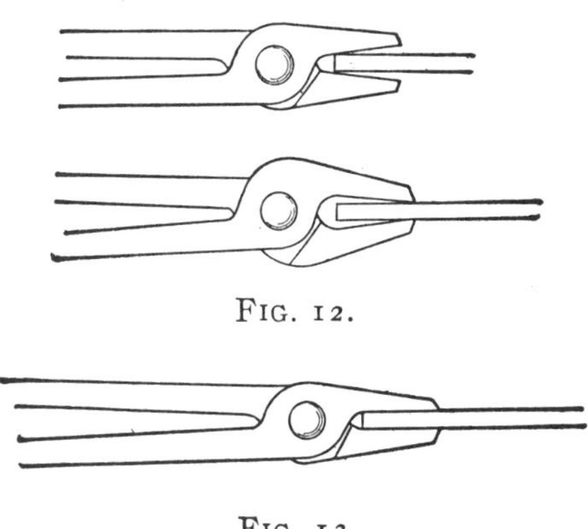

FIG. 12.

FIG. 13.

When properly fitted, the jaws should touch the work the entire length, as illustrated in Fig. 13. With properly fitted tongs the work may be held firmly, but if fitted as shown in Fig. 12 there is always a very "wobbly" action between the jaws and the work.

To fit a pair of tongs to a piece of work, the jaws should be heated red-hot, the piece to be held placed between them, and the jaws closed down tight around the piece with a hammer. To prevent the handles from being brought too close together while the tongs are being fitted, a short piece of stock should be held between them just back of the jaws. If the handles are too far apart, a few blows just back of the eye will close them up.

Flatter—Set-hammer—Swage—Fuller—Swage-block.—Among the commonest tools used in forge-work are the ones mentioned above.

The flatter, Fig. 14, as its name implies, is used for flattening and smoothing straight surfaces.

The face of the flatter is generally from 2 inches to 3 inches square, and should be kept perfectly smooth with the edges slightly rounding.

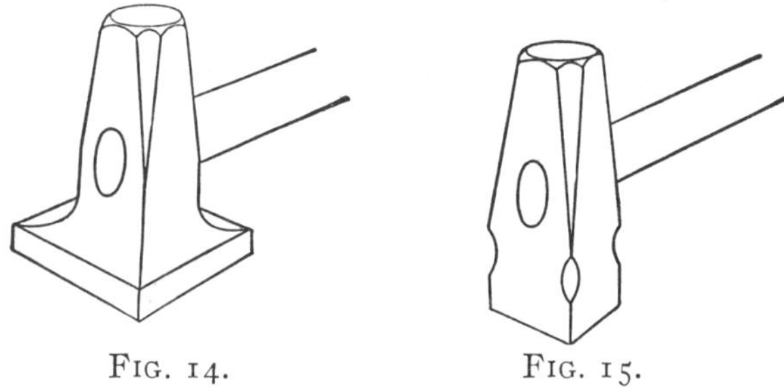

FIG. 14. FIG. 15.

Fig. 15 shows a set-hammer. This is used for finishing parts which cannot be reached with the flatter, up into corners, and work of that character. The face of this tool also should be smooth and flat, with the corners more or less rounded, depending on the work it is intended to do.

Set-hammers for small work should be about 1 or $1\frac{1}{4}$ inches square on the face.

"Set-hammer" is a name which is sometimes given to almost any tool provided with a handle, which tool in use is held in place and struck with another hammer. Thus, flatters, swages, fullers, etc., are sometimes classed under the general name of set-hammers.

Fullers, Fig. 16, are used for finishing up filleted corners, forming grooves, and for numerous purposes which will be given more in detail later.

They are made in a variety of sizes, the size being determined by the shape of the edge *A*. On a ½ inch fuller this edge would be a half-circle ½ inch in diameter; on a ¾ inch it would be ¾ inch in diameter, etc.

Fullers are made "top" and "bottom." The one shown with a handle is a "top" fuller, and the lower one in the illustration is a "bottom" fuller and has the stem forged to fit into the square hardie-hole of the anvil. This stem should be a loose fit in the hardie-hole. Tools of this character should never be used on an anvil where they fit so tightly that it is necessary to drive them into place.

A top and bottom swage is shown in Fig. 17. The swages shown here are for finishing round

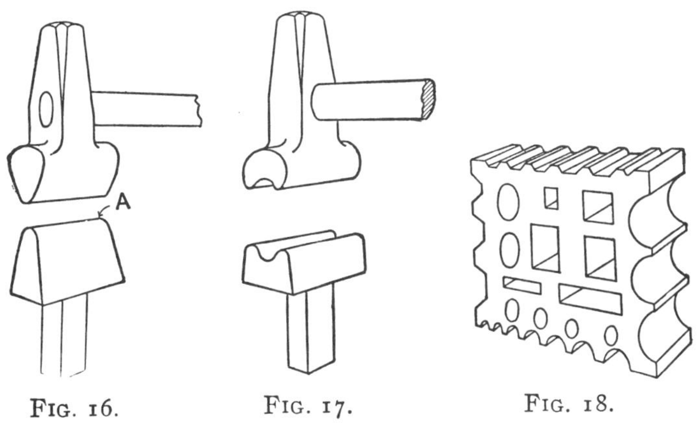

FIG. 16. FIG. 17. FIG. 18.

work; but swages are made to be used for other shapes as well.

Swages are sized according to the shape they are made to fit. A 1-inch round swage, for instance, is made to fit a circle 1 inch in diameter, and would be used for finishing work of that size.

All of the above tools are made of low-carbon tool steel.

A swage-block is shown in Fig. 18. These blocks are made in a variety of shapes; the illustration showing a common form for general use. This block is made of cast iron and is about $3\frac{1}{2}$ inches thick. It has a wide range of uses and is very convenient for general work, where it takes the place of a good many special tools.

CHAPTER II.

WELDING.

Welding-heat.—A piece of wrought iron or mild steel, when heated, as the temperature increases becomes softer and softer until at last a heat is reached at which the iron is so soft that if another piece of iron heated to the same point touches it, the two will stick together. The heat at which the two pieces will stick together is known as the welding-heat. If the iron is heated much beyond this point, it will burn. All metals cannot be welded (in the sense in which the term is ordinarily used). Some, when heated, remain very dense and retain almost their initial hardness until a certain heat is reached, when a very slight rise of temperature will cause them to either crumble or melt. Only those metals which, as the temperature is increased, become gradually softer, passing slowly from the solid to the liquid state, can be welded easily. Metals of this kind just before melting become soft and more or less pasty, and it is in this condition that they are most weldable. The greater the range of temperature through which the metal remains pasty the more easily may it be welded.

In nearly all welding the greatest trouble is in heating the metal properly. The fire must be clean and bright or the result will be a "dirty" heat; that is, small pieces of cinder and other dirt will stick to the metal, get in between the two pieces, and make a bad weld.

Too much care cannot be used in welding; if the pieces are too cold they will not stick, and no amount of hammering will weld them. On the other hand, if they are kept in the fire too long and heated to too high a temperature they will be burned, and burned iron is absolutely worthless.

The heating must be done slowly enough to insure the work heating evenly all the way through. If heated too rapidly, the outside may be at the proper heat while the interior metal is much colder; and, as soon as taken from the fire, this cooler metal on the inside and the air almost instantly cool the surface to be welded below the welding temperature, and it will be too cold to weld by the time any work can be done on it.

If the pieces are properly heated (when welding wrought iron or mild steel), they will feel sticky when brought in contact.

When welding, it is best to be sure that everything is ready before the iron is taken from the fire. All the tools should be so placed that they may be picked up without looking to see where they are. The face of the anvil should be perfectly clean, and the hammer in such a position that it will not be knocked out of the way when the work is placed on the anvil for welding.

All the tools being in place, and the iron brought to the proper heat, the tongs should be held in such a way that the pieces can be easily placed in position for welding without changing the grip or letting go of them; then, when everything is ready, the blast should be shut off, the pieces taken from the fire, placed together on the anvil, and welded together with rapid blows of the hammer, welding (after the pieces are once stuck together) the thin parts first, as these are the parts which naturally cool the quickest.

Burning Iron or Steel.—The statement that iron can be burned seems to the beginner to be rather exaggerated. The truth of this can, however, be very easily shown. If a bar of iron be heated in the forge and considerable blast turned on, the bar will grow hotter and hotter, until at last sparks will be seen coming from the fire. These sparks, which are quite unlike the ordinary ones from the fire, are white and seem to explode and form little white stars.

These sparks are small particles of burning iron which have been blown upward out of the fire.

The same sparks may be made by dropping fine iron-filings into a gas-flame, or by burning a piece of oily waste which has been used for wiping up iron-filings.

If the bar of iron be taken from the fire at the time these sparks appear, the end of the bar will seem white and sparkling, with sparks, like stars similar to those in the fire, coming from it. If the heating be continued long enough, the end of the bar will be partly consumed, forming lumps similar to the "clinkers" taken from a coal fire.

To burn iron two things are necessary: a high enough heat, and the presence of oxygen.

As noted before, when welding, care must be taken not to have too much air going through the fire; in other words, not to have an oxidizing fire.

If the fire is not an oxidizing one, there is not so much danger of injury to the iron by burning and the forming of scale.

Iron which has been overheated and partially burned has a rough, spongy appearance and is brittle and crumbly.

Use of Fluxes in Welding.—When a piece of iron or steel is heated for welding under ordinary conditions the outside is oxidized; that is, a thin film of iron oxide is formed. This oxide is the black scale which is continually falling from heated iron and is formed when heated iron is brought in contact with the air. This oxide of iron is not fluid except at a very high heat, and, if allowed to stay on the iron, will prevent a good weld.

When welding without a flux the iron is brought to a high enough heat to melt the oxide, which is forced from between the welding pieces by the blows of the hammer.

This heat may easily be taken when welding ordinary iron; but when working with some machine-steel, and particularly tool-steel, the metal cannot be heated to a high enough temperature to melt the oxide without burning the steel.

From the above it would seem impossible to weld steel, as it cannot be heated under ordinary conditions without oxidizing, but by the use of a flux this difficulty may be overcome and the oxide melted at a lower temperature.

The flux (sand and borax are the most common) should be sprinkled on the part of the piece to be welded when it has reached about a yellow heat, and the heating continued until the metal is at a proper temperature to be soft enough to weld, but care should be taken to see that the flux covers the parts to be welded together.

The flux has a double action; in the first place, as it melts it flows over the piece and forms a protecting covering which prevents oxidation, and also when raised to the proper heat dissolves the oxide that has already formed.

The oxide melts at a much lower heat when combined with the flux than without it, and to melt the oxide is the principal use of the flux. The metal when heated in contact with the flux becomes soft and "weldable" at a lower temperature than when without it.

Ordinary borax contains water which causes it to bubble up when heated. If the heating is continued at a high temperature, the borax melts and runs like water; this melted borax, when cooled, is called borax-glass.

Borax for welding is sometimes fused as above and then powdered for use.

Sal ammoniac mixed with borax seems to clean the surface better than borax alone. A flux made of one part of sal ammoniac and four parts borax works well, particularly when welding tool-steel, and is a little better than borax alone.

Most patented welding compounds have borax as a basis, and are very little, if any, better than the ordinary mixture given above.

The flux does not in any way stick the pieces together or act as a cement or glue. Its use is principally to help melt the oxide already formed and to prevent the formation of more.

Iron filings are sometimes mixed with borax and used as a flux.

When using a flux the work should always be scarfed the same as when no flux is used. The pieces can be welded, however, at a lower heat.

Fagot or Pile Welding.—When a large forging is to be made of wrought iron, small pieces of "scrap" iron (old horseshoes, bolts, nuts, etc.) are placed together in a square or rectangular pile on a board, bound together with wire, heated to welding heat in a furnace, and welded together into one solid lump, and the forging made from this. If there is not enough metal in one lump, several are made in this way and afterward welded to each other—making one large piece. This is known as fagot or pile welding.

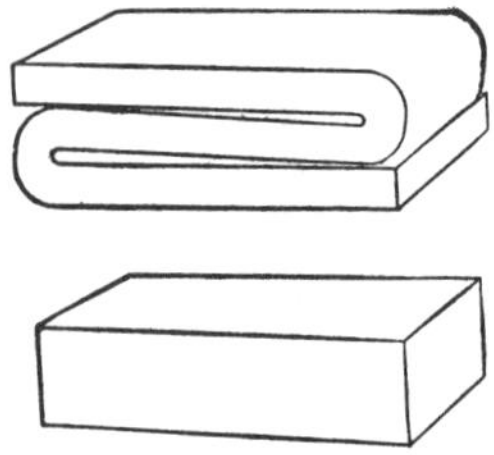

FIG. 19.

Sand is used for fluxing to a large extent on work of this kind.

Sometimes a small fagot weld is made by laying two or more pieces together and welding them their entire length, or one piece may be doubled together several times and welded into a lump. Such a weld is shown in Fig. 19, which shows the piece before welding and also after being welded and shaped.

Scarfing.—In a fagot weld the pieces are not prepared or shaped for each other, being simply laid together and welded, but for most welding the ends of the pieces to be joined should be so shaped that they will fit together and form a smooth joint when welded. This is called scarfing. It is very important that the scarfing be properly done, as a badly shaped scarf will probably spoil the weld. For instance, if an attempt be made to weld two bars together simply by overlapping their ends, as in Fig. 20, the weld when finished would be something like Fig. 21. Each bar would

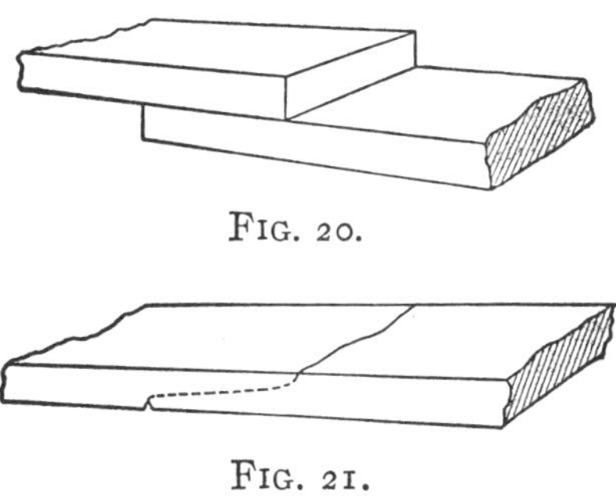

FIG. 20.

FIG. 21.

be forged into the other and leave a small crack where the end came. On the other hand, if the ends of the bars were properly scarfed or pointed, they could be welded together and leave no mark—making a smooth joint.

Lap-welds are sometimes made without scarfing when manufacturing many pieces alike, but this should not be attempted in ordinary work.

Lap-weld Scarf.—In preparing for the lap-weld, the ends of the pieces to be welded should be first upset until they are considerably thicker than the rest of the bar. This is done to allow for the iron which burns off, or is lost by scaling, and also to allow for the hammering which must be done when welding the pieces together. To make a proper weld the joint should be well hammered together, and as this reduces the size of the iron at that point the pieces must be upset to allow for this reduction in size in order to have the weld the same size as the bar.

If the ends are not upset enough in the first place, it requires considerable hard work to upset the weld after they are joined together. Too much upsetting does no harm, and the extra metal is very easily worked into shape. To be on the safe side it is better to upset a little more than is absolutely necessary—it may save considerable work afterwards.

If more than one heating will be necessary to make a weld, the iron should be upset just that much more to allow for the extra waste due to the second or third heating.

Flat Lap-weld.—The lap-weld is the weld ordinarily used to join flat or round bars of iron together end to end.

Following is a description of a flat lap-weld: The ends of the pieces to be joined must be first upset. When heating for upsetting heat only the end as shown in Fig. 22, where the shaded part indicates the hotter metal. To heat this way place only the extreme *end* of the bar in the fire, so the heat will not run back too far. The end should be upset until it looks about like Fig. 23.

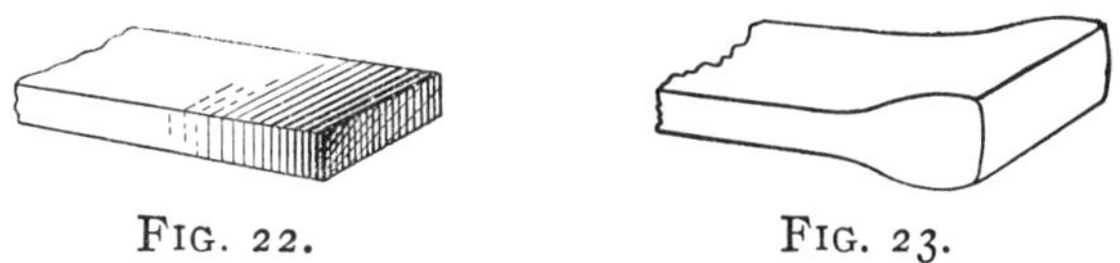

FIG. 22. FIG. 23.

When starting to shape the scarf use the round or pene end of the hammer. Do not strike directly *down* on the work, but let the blows come at an angle of about 45 degrees and in such a way as to force the metal back toward the base as shown in Fig.

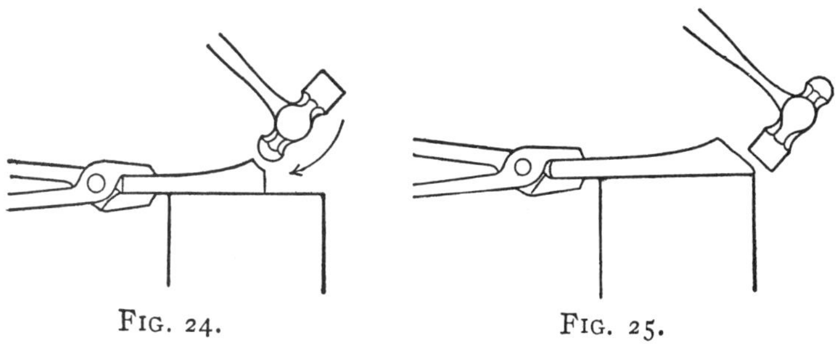

FIG. 24. FIG. 25.

24. This drives the metal back and makes a sort of thick ridge at the beginning of the scarf. In finishing the scarf, use the flat face of the hammer, and bring the piece to the very edge of the anvil, as in this way a hard blow may be struck without danger of hitting the anvil instead of the work. The proper position is shown in Fig. 25.

The scarfs should be shaped as in Fig. 26, leaving them slightly convex and not concave, as shown in Fig. 27.

The reason for this is that if the scarfed ends be concave when the two pieces are put together, a

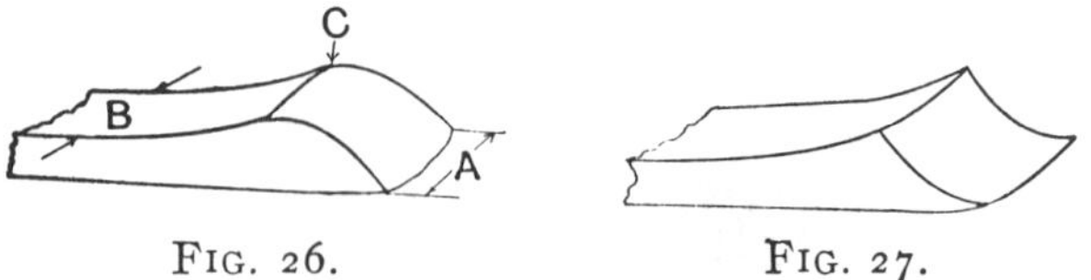

FIG. 26. FIG. 27.

small pocket or hollow will be left between them, the scarfs touching only the edges. When the weld is hammered together, these edges being in contact will naturally weld first, closing up all outlet to the pocket. As the surface of the scarf is more or less covered with melted scale and other impurities, some of this will be held in the pocket and make a bad place in the weld. On the other hand, if the scarfs are convex, the metal will first stick in the very center of the scarf, forcing out the melted scale at the sides of the joint as the hammering continues.

The length of the scarf should be about $1\frac{1}{2}$ times the thickness of the bar; thus on a bar $\frac{1}{2}''$ thick the scarf should be about $\frac{3}{4}''$ long.

The width of the end *A* should be slightly less than the width *B* of the bar. In welding the two pieces together, the first piece should be placed scarf side up on the anvil, and the second piece laid on top, scarf side down, in such a way that the thin edges of the second piece will lap over the thick ridge *C* on the first piece as shown in Fig. 28. The piece which is laid on top should be held by the smith doing the welding, the other may be handled by the helper.

The helper should place his piece in position on the anvil first. As it is rather hard to lay the other piece directly on top of this and place it exactly in the right position, it is better to rest the second piece on the corner of the anvil as

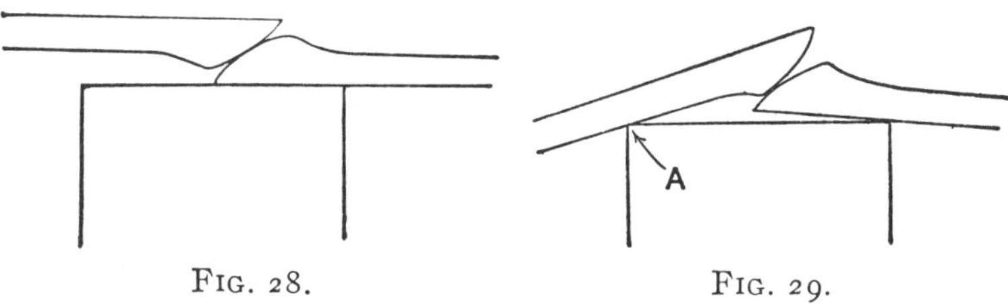

FIG. 28. FIG. 29.

shown at *A*, Fig. 29, and thus guide it into position. In this way the piece may be steadied and placed on the other in the right position without any loss of time.

When heating for a lap-weld, or for that matter any weld where two pieces are joined together, great care should be taken to bring both pieces to the same heat at the same time. If one piece heats faster than the other, it should be taken from the fire and allowed to cool until the other piece "catches up" with it. It requires some practice to so place the pieces in the fire that they will be heated uniformly and equally. The tips particularly must be watched, and it may be necessary to cool them from time to time in the water-bucket to prevent the extreme ends from burning off.

The fire *must* be clean, and the heating should be done slowly in order to insure its being done evenly.

Just before taking the pieces from the fi-- they

should be turned scarf side down for a short time, to be sure that the surfaces to be joined will be hot.

More blast should be used at the last moment than when starting to heat.

The only way to know how this heating is going on is to take the pieces from the fire from time to time and look at them. The color grows lighter as the temperature increases, until finally, when the welding heat is reached, the iron will seem almost white. The exact heat can only be learned by experience; but the workman should recognize it after a little practice as soon as he sees it.

To get an indication of the heat, which will help sometimes, watch the sparks that come from the fire. When the little, white, explosive sparks come they show that some of the iron has been heated hot enough to be melted off in small particles and is burning. This serves as a rough indication that the iron is somewhere near the welding heat. This should never be relied on entirely, as the condition of the fire has much to do with their appearance.

Round Lap-weld.—The round lap-weld—the weld used to join round bars end to end—is made in much the same way as the ordinary or flat lap-weld. The directions given for making the flat weld apply to the round lap as well, excepting that the scarf is slightly different in shape. The proper shape of scarf is shown in Fig. 30, which gives the top and side views of the piece. One side is

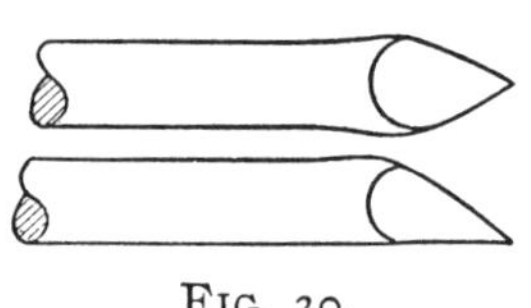

FIG. 30.

left straight, the other three sides tapering in to meet it in a point. The length of the scarf should be about one and one-half times the diameter of the bar. Always be sure, particularly in small work, that the pieces are scarfed to a *point*, and not merely flattened out. The greatest difficulty with this weld is to have the points of the pieces well welded, as they cool very rapidly after leaving the fire. The first blows, after sticking the pieces together, should cover the points. The weld should be made *square* at first and then rounded. The weld is not so apt to split while being hammered if welded square and then worked round, as it would be if hammered round at first.

If the scarf were made wide on the end like the ordinary lap-weld, it would be necessary to hammer clear around the bar in order to close down the weld; but with the pointed scarf, one blow on each point will stick the work in place, making it much more quickly handled.

Ring Weld, Round Stock. A ring formed from round stock may be made in two ways; that is, by scarfing before or after bending into shape. When scarfed *before* bending, the length of stock should be carefully calculated, a small amount

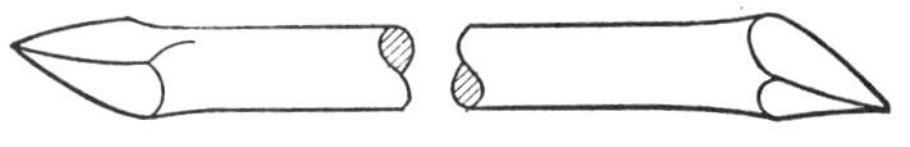

FIG. 31.

being added for welding, and the ends upset and scarfed exactly the same as for a round lap-weld.

Care should be taken to see that the scarfs come on opposite sides of the piece.

Fig. 31 shows a piece of stock scarfed ready for bending.

After scarfing, the piece should be bent into a ring and welded, care being taken when bending to see that the points of the scarf lie as indicated at *A*, Fig. 32, and not as shown at *B*.

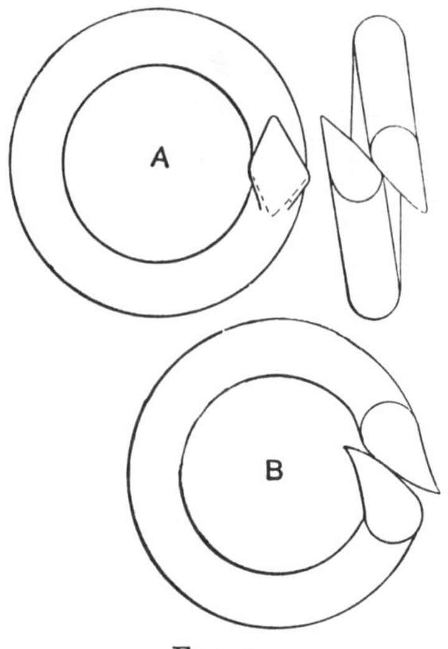

FIG. 32.

When the points of the scarfs are lapped as shown at *A*, most of the welding may be done while the ring lies flat on the anvil, the shaping being finished over the horn. If the points are lapped the other way, *B*, the welding also must be done over the horn, making it much more awkward to handle.

The second way of welding the ring is practically the same as that of making a chain link, and the same description of scarfing will answer for both, the stock being cut and bent into a ring, with the ends a little distance apart; these ends are then scarfed the same as described below for a link scarf and welded in exactly the same manner as described for making the other ring.

Chain-making.—The first step in making a link is to bend the iron into a U-shaped piece, being careful to keep the legs of the U exactly even in length. The piece should be gripped at the lower end of the U, the two ends brought to a high heat, scarfed, bent into shape together, reheated, and welded.

To scarf the piece place one end of the U on the anvil, as shown in Fig. 33, and strike one blow on it; move it a short distance in the direction shown by the arrow and strike another blow. This should be continued until the edge or corner of the piece is reached, moving it after each blow.

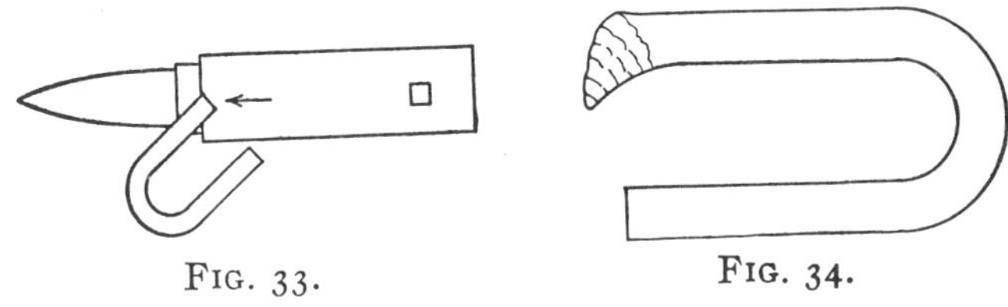

FIG. 33. FIG. 34.

This operation leaves a series of little steps on the end of the piece, and works it out in a more or less pointed shape, as shown in Fig. 34.

This scarf may be finished by being brought more to a point by a few blows over the horn of the anvil. The ends should then be bent together and welded. Fig. 35 shows the steps in making the link and two views of the finished link. The link is sometimes left slightly thicker through the weld. A second link is made—all but welding—spread open, and the first link put on it, closed up again, and welded. A third is joined to this etc.

When made on a commercial scale, the links are

not scarfed but bent together and welded in one heat.

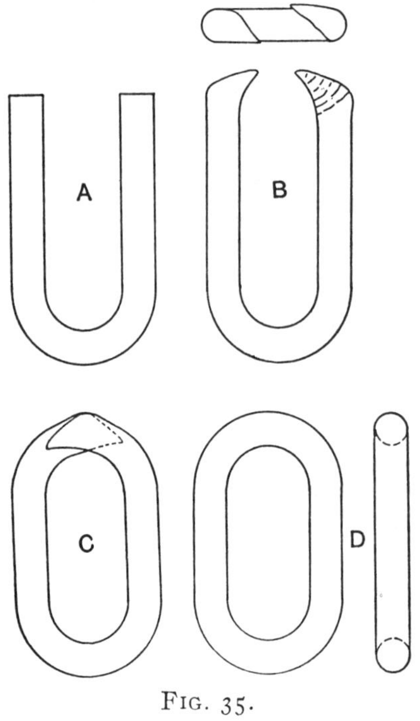

FIG. 35.

Ring, or Band.—A method of making a ring from

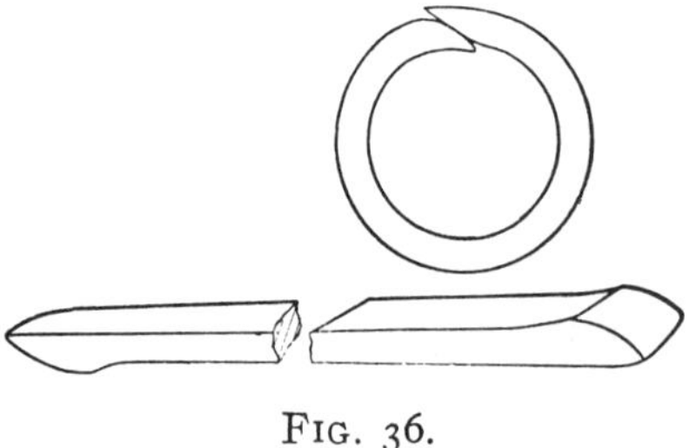

FIG. 36.

flat iron is shown in Fig. 36, which shows the stock before and after bending into shape.

The stock is cut to the correct length, upset, and scarfed exactly the same as for a flat lap-weld. The piece is bent into shape and welded over the horn of the anvil. The ring must be heated for welding very carefully or the outside lap will burn before the inside is hot enough to weld.

In scarfing this—as in making other rings—care must be taken to have the scarfs come on opposite sides of the tock.

Washer, or Flat Ring. — In this weld flat stock is used bent edgewise into a ring without any preparation. The corners of the ends are trimmed off parallel after the stock is bent as shown in Fig. 37.

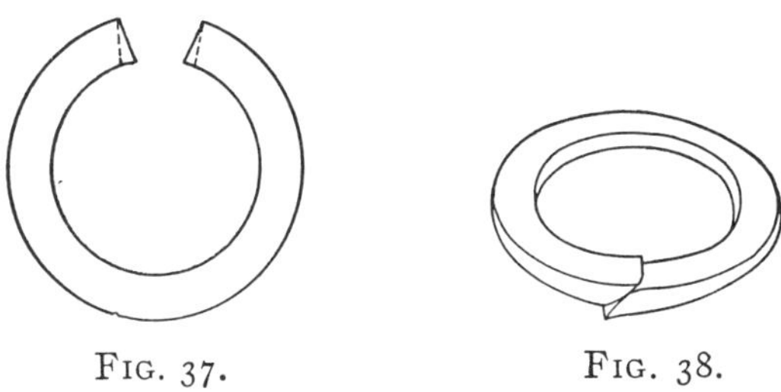

Fig. 37. Fig. 38.

After trimming the ends are scarfed with a fuller or pene end of a hammer and lapped ready for welding (Fig. 38).

When heating for welding, the ring should be turned over several times to insure uniformity in heating.

If the work is particular, the ends of the stock should be upset somewhat before bending into shape.

Butt-weld. — This is a weld where the pieces are butted together without any slanting scarfs, leaving a square joint through the weld.

When two pieces are so welded the ends should be slightly rounded, simillar to Fig. 39, which shows two pieces ready for welding. If the ends are convex as shown, the scale and other impurity sticking to the metal is forced out of the joint. If the ends were concave this matter would be

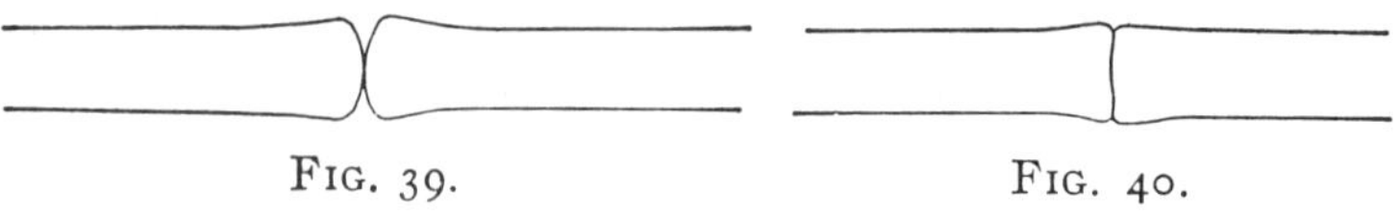

FIG. 39. FIG. 40.

held between the pieces and make a poor weld. The pieces are welded by being struck on the ends and driven together. This, of course, upsets the metal near the weld and leaves the piece something like Fig. 40, showing a slight seam where the rounded edges of the ends join. This upset part is worked down to size at a welding heat, leaving the bar smooth.

A butt-weld is not as safe or as strong as a lap-weld.

When the pieces are long enough they may be welded right in the fire. This is done by placing the pieces in the fire in the proper position for welding; a heavy weight is held against the projecting end of one piece—to "back it up"—and the weld is made by driving the pieces together by hammering on the projecting end of the second piece. As soon as the work is "stuck," the weld is taken from the fire and finished on the anvil.

Jump Weld.—Another form of butt-weld, Fig. 41, is the "jump" weld, which, however, is a form which should be avoided as much as possible, as it is very liable to be weak. When making a weld like this, the piece which is to be "jumped," or "butted," on to the other piece should have its end upset in such a way as to flare out and form a sort of flange the wider the better. When the weld is made, this flange—indicated by the arrow—can be welded down with a hammer, or set-hammers, and make a fairly strong weld.

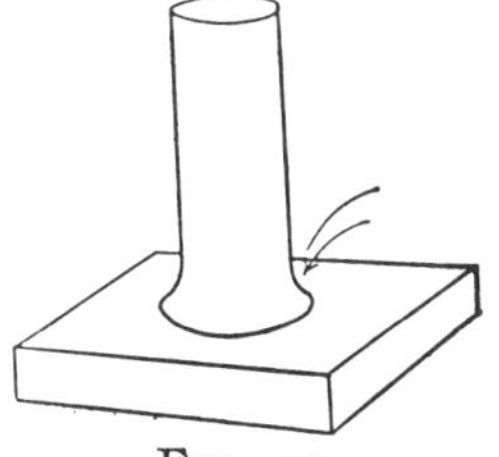

FIG. 41.

Split Weld; Weld for very Thin Steel.—Very thin stock is sometimes difficult to join with the ordinary lap-weld for the reason that the stock is so thin that if the pieces are taken from the fire at the proper heat they will be too cold to weld before they can be properly placed together on the anvil.

This difficulty is somewhat overcome by scarfing the ends, similar to Fig. 42. The ends are tapered

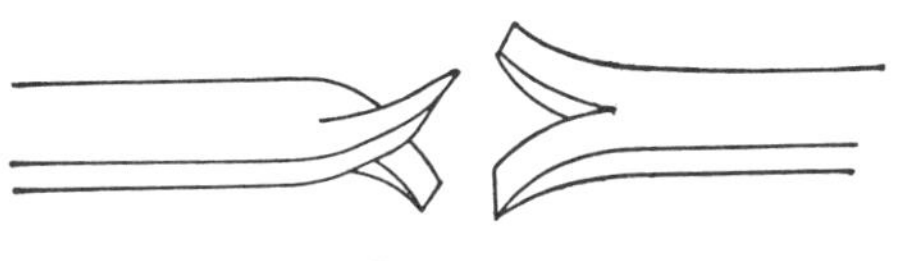

FIG. 42.

to a blunt edge and split down the center for half an inch or so, depending on the thickness of stock. One half of each split end is bent up, the other down; the ends are pushed tightly together and the split parts closed down on each other, as shown

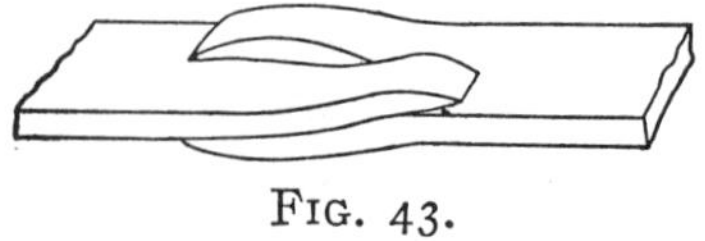

FIG. 43.

in Fig. 43. The joint may then be heated and welded.

This is a weld sometimes used for welding spring steel, or iron to steel.

Split Weld; Heavier Stock.—A *split weld* for heavier stock is shown ready for welding in Fig.

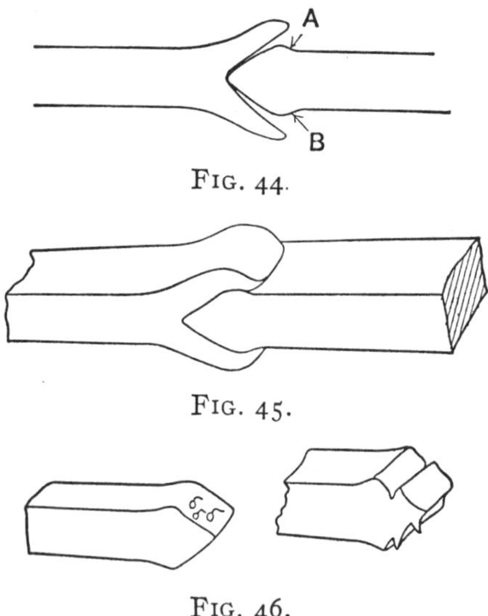

FIG. 44.

FIG. 45.

FIG. 46.

45, Fig. 44 showing the two pieces before they are put together. In this weld the ends of the pieces are first upset and then scarfed, one piece being split and shaped into a Y, while the other has its end brought to a point with the sides of the bar just back of the point bulging out slightly as shown at *A* and *B*. This bulge is to prevent the two pieces from slipping apart.

When properly shaped the two pieces are driven together and the sides, or lips, of the Y-shaped scarf closed down over the pointed end of the other piece. The lips of the Y should be long enough to lap over the bulge on the end of the other piece and thus prevent the two pieces from slipping apart. The pieces are then heated and welded. Care must be taken to heat slowly, that the pointed part may be brought to a welding heat without burning the outside piece. Borax, sand, or some other flux should be used. (Sometimes the faces of the scarfs are roughened or notched with a chisel, as shown in Fig. 46, to prevent the pieces from slipping apart.)

This is the weld that is often used when welding tool-steel to iron or mild steel.

Sometimes the pieces are heated separately to a welding heat before being placed together. Good results may be obtained this way when tool-steel is welded to iron or mild steel, as the tool-steel welds at a much lower temperature than either wrought iron or mild steel, and if the two pieces are heated separately, the other metal may be raised to a much higher temperature than the tool-steel.

Angle Weld.—In all welding it should be remembered that the object of scarfing is to so shape the pieces to be welded that they will *fit* together and form a smooth joint when properly hammered.

Frequently there are several equally good methods of scarfing for the same sort of weld, and it should be remembered that the method given here is not necessarily the only way in which the particular weld can be made.

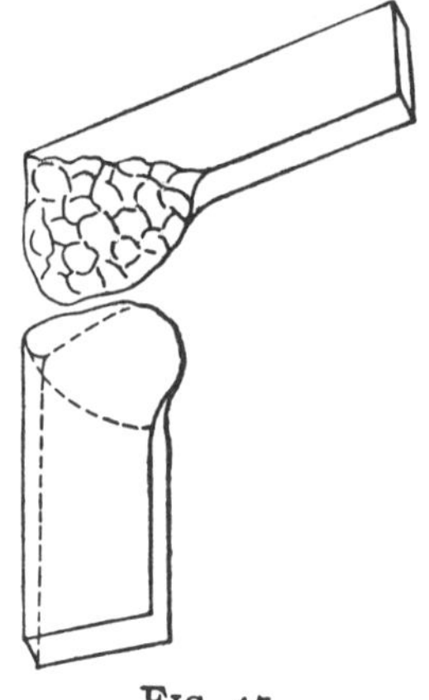

FIG. 47.

Fig. 47 shows one way of scarfing for a right-angle weld made of flat iron. Both pieces are scarfed exactly alike. The scarfing is done with the pene end of the hammer. If necessary the ends of the pieces may be upset before scarfing.

As in all other welds, care must be taken to so shape the scarfs that when they are placed together they will touch in the center, and not around the edges, thus leaving an opening for forcing out the impurities which collect on the surfaces to be welded.

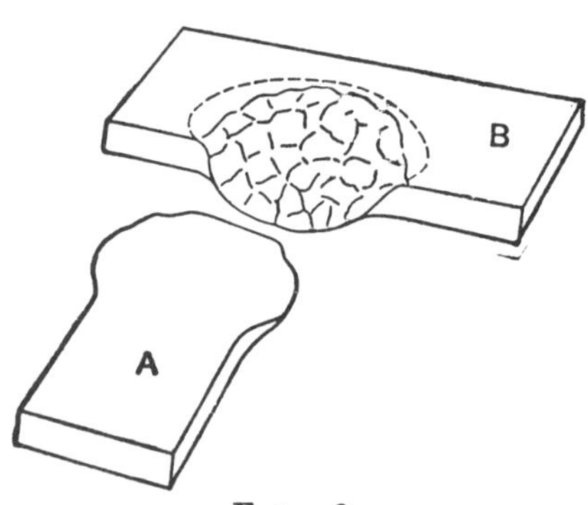

FIG. 48.

"T" Weld.—A method of scarfing for a "T" weld is illustrated in Fig. 48.

The stem, *A*, should be placed on the bar, *B*, when welding in about the position shown by the dotted line on *B*.

"T" Weld, Round Stock.—Two methods of scarfing for a "T" weld made from round stock are shown in Fig. 49.

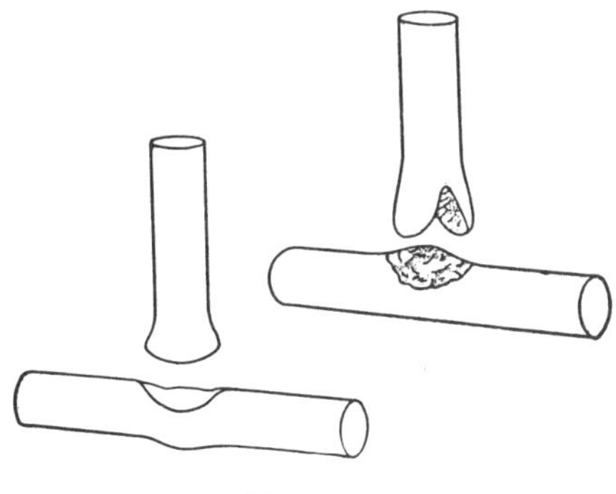

FIG. 49.

The scarfs are formed mostly with the pene end of the hammer.

The illustration will explain itself. The stock should be well upset in either method.

Welding Tool-steel.—The general method of scarfing is the same in all welding; but when tool-steel is to be welded, either to itself or to wrought iron or mild steel, more care must be used in the heating than when working with the softer metals alone.

The proper heat for welding tool-steel—about a bright yellow—can only be learned by experiment. If the tool-steel is heated until the sparks fly, a light blow of the hammer will cause it to crumble and fall to pieces.

When welding mild steel or wrought iron to tool-steel, the tool-steel should be at a lower heat than the other metal, which should be heated to its regular welding heat.

The flux used should be a mixture of about one part sal ammoniac and four parts borax.

Tool-steel of high carbon, and such as is used for files, small lathe tools, etc., can seldom be welded to itself in a satisfactory manner. What appears to be a first-class weld may be made, and the steel may work up into shape and seem perfect—may, in fact, be machined and finished without showing any signs of the weld—but when the work is hardened, the weld is almost certain to crack open.

Spring steel, a lower carbon steel, may be satisfactorily welded if great care be used.

CHAPTER III.

CALCULATION OF STOCK FOR BENT SHAPES.

Calculating for Angles and Simple Bends. — It is often necessary to cut the stock for a forging as nearly as possible to the exact length needed. This length can generally be easily obtained by measment or calculation.

About the simplest case for calculation is a plain right-angle bend, of which the piece in Fig. 50 will serve as an example.

This piece as shown is a simple right-angle bend made from stock 1″ through, 8″ long on the outside of each leg.

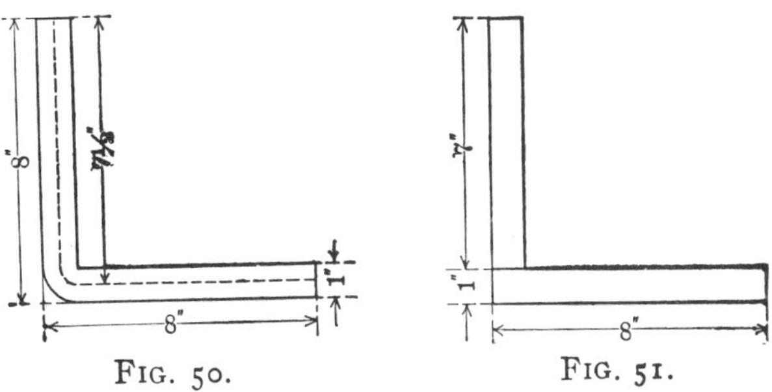

FIG. 50. FIG. 51.

Suppose this to be made of wood in place of iron. It is easily seen that a piece of stock 1″ thick and 15″ long would make the angle by cutting off 7″

from one end and fastening this piece to the end of the 8″ piece, as shown in Fig. 51.

This is practically what is done when the angle is made of iron—only, in place of cutting and fastening, the bar is bent and hammered into shape.

In other words, any method which will give the length of stock required to make a shape of uniform section in wood, if no allowance is made for cutting or waste, will also give the length required to make the same shape with iron.

An easier way—which will serve for calculating lengths of all bent shapes—is to measure the length of an imaginary line drawn through the center of the stock. Thus, if a dotted line should be drawn through the center of stock in Fig. 50, the length of each leg of this line would be $7\frac{1}{2}$″, and the length of stock required 15″, as found before.

No matter what the shape when the stock is left of uniform width through its length, this length of straight stock may always be found by measuring the length of the center line on the bent shape. This may be clearly shown by the following experiment.

Experiment to Determine Part of Stock which Remains Constant in Length while Bending.—Suppose a straight bar of iron with square ends be taken and bent into the shape shown in Fig. 52. If the length of the bar be measured on the inside edge of the bend and then on the outside, it will be found that the inside length is considerably shorter

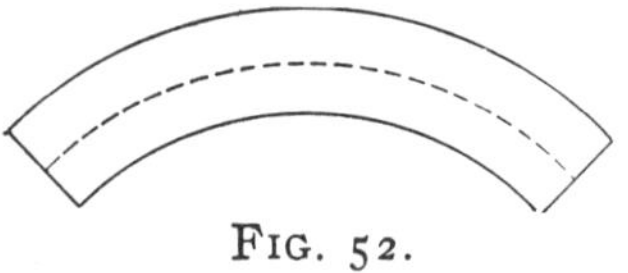

FIG. 52.

than the outside; and not only this, but the inside will be shorter than the original bar, while the outside will be longer. The metal must therefore squeeze together or upset on the inside and stretch or draw out on the outside. If this is the case, as it is, there must be some part of the bar which when it is bent neither squeezes together nor draws out, but retains its original length, and this part of the bar lies almost exactly in the center, as shown by the dotted line. It is on this line of the bent bar that the measuring must be done in order to determine the original length of the straight stock, for this is the only part of the stock which remains unaltered in length when the bar is bent.

To make the explanation a little clearer, suppose a bar of iron is taken, polished on one side, and lines scratched upon the surface, as shown in the lower drawing of Fig. 53, and this bar then bent into the shape shown in the upper drawing. Now if the length of each one of these lines be measured and the measurements compared with the length of the same lines before the bar was bent, it would be found that the line *AA*, on the outside of the bar, had lengthened considerably; the line *BB* would be somewhat lengthened, but not as much as *AA*; and *CC* would be lengthened less than *BB*. The line *OO*, through the center of the bar, would measure almost exactly the same as when the bar was straight. The line *DD* would be found to be shorter than *OO* and *FF* shorter than any other. The line *OO*, at the center of the bar, does not change its length when the bar is bent; consequently, to determine the length of straight stock required to bend into any shape, measure the

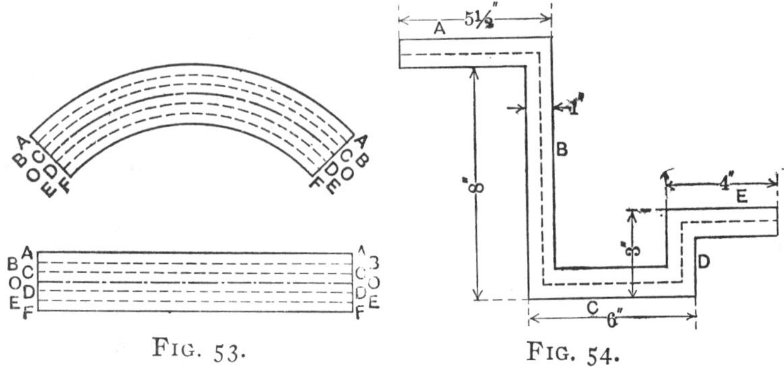

FIG. 53. FIG. 54.

length of the line following the center of the stock of the bent shape.

As another example Fig. 54 will serve.

Suppose a center line be drawn, as shown by the dotted line. As the stock is 1" thick, the length of the center line of the part *A* will be 5", at *B* 8", *C* 5", *D* 2", *E* 3½", and the total length of stock required 21½".

A convenient form for making calculations is as follows:

A = 5"
B = 8"
C = 5"
E = 3½"

Total. . . 21½" = length of stock required.

Curves. Circles. Methods of Measuring.—On circles and curves there are several different methods which may be employed in determining the length of stock, but the same principle must be followed in any case—the length must be measured along the center line of the stock.

One way of measuring is to lay off the work full size. On this full-size drawing lay a string or thin, easily bent wire in such a way that it follows the shape of the bend through its entire length, being careful that the string is laid along the center of the stock.

The string or wire may then be straightened and the length measured directly.

Irregular shapes or scrolls are easily measured in this way.

Another method of measuring stock for scrolls, etc., is to step around a scroll with a pair of dividers with the points a short distance apart, and then lay off the same number of spaces in a straight line and measure the length of that line. This is of more use in the drawing-room than in the shop.

Measuring-wheel.—Still another way of measuring directly from the drawing is to use a light measuring-wheel, similar to the one shown in Fig. 55, mounted in some sort of a handle. This is a thin light wheel generally made with a circumference of about 24″. The side of the rim is sometimes graduated in inches by eighths. To use it, the wheel is placed lightly in contact with the line or object which it is wished to measure, with the zero-mark on the wheel corresponding to the point from which the measurement is started. The wheel is then

FIG. 55.

pushed along the surface following the line to be measured, with just pressure enough to make it revolve. By counting the revolutions made and setting the pointer or making a mark on the wheel to correspond to the end of the line when it is reached, it is an easy matter to push the wheel over a straight line for the same number of revolutions and part of a revolution as shown by the pointer and measure the length. If the wheel is graduated, the length run over can of course be read directly from the figures on the side of the wheel.

Calculating Stock for Circles.—On circles and parts of circles, the length may be calculated mathematically, and in the majority of cases this is probably the easiest and most accurate method. This is done in the following way: The circumference, or distance around a circle, is equal to the diameter multiplied by $3\frac{1}{7}$ (or more accurately, 3.1416).

As an illustration, the length of stock required to bend up the ring in Fig. 56 is calculated as follows: The inside diameter of the ring is 6″ and the stock 1″ in diameter. The length must, of course, be measured along the center of the stock, as shown by the dotted line. It is the diameter of this circle, made by the dotted line, that is used for calculating the length of stock; and for convenience this may be called the "calculating" diameter, shown by *C* in Fig. 56.

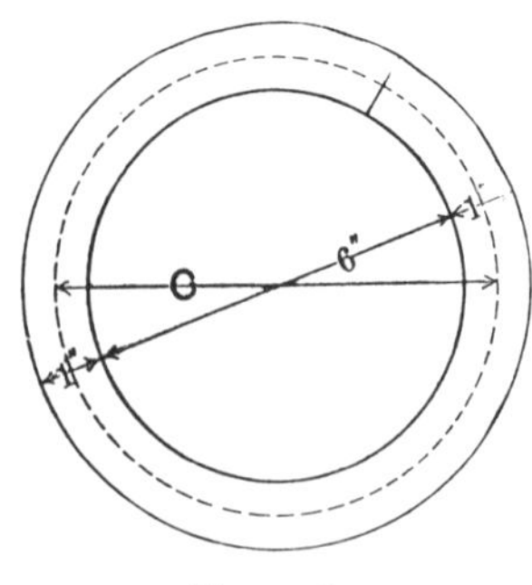

FIG. 56.

The length of this calculating diameter is equal to the inside diameter of the ring with one-half the thickness of stock added at each end, and in this case would be $\frac{1}{2}'' + 6'' + \frac{1}{2}'' = 7''$.

The length of stock required to make the ring would be $7'' \times 3\frac{1}{7} = 22''$; or, in other words, to find the length of stock required to make a ring, multiply the diameter of the ring, measured from center to center of the stock, by $3\frac{1}{7}$.

Calculating Stock for "U's."—Some shapes may be divided up into straight lines and parts of circles and then easily calculated. Thus the U shape in Fig. 57 may be divided into two straight sides and a half-circle end. The end is half of a circle having an outside diameter of $3\frac{1}{2}''$. The calculating diameter of this circle would be $3''$, and the length of stock required for an entire circle this size $3 \times 3\frac{1}{7} = 9\frac{3}{7}$, which for convenience we may call $9\frac{3}{8}''$, as this is near enough for ordinary work. As the forging calls for only half a circle, the length needed would be $9\frac{3}{8}'' \div \frac{1}{2} = 4\frac{11}{16}''$.

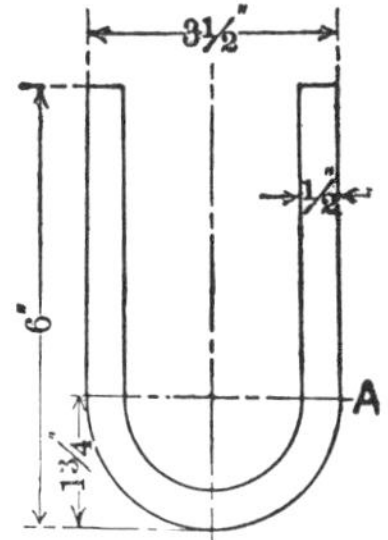

FIG. 57.

As the circle is $3\frac{1}{2}''$ outside diameter, half of this diameter, or $1\frac{3}{4}''$, must be taken from the total length of the U to give the length of the straight part of the sides; in other words, the distance from the line *A* to the extreme end of the U is half the diameter of the circle, or $1\frac{3}{4}''$. This leaves the straight sides each $4\frac{1}{4}''$ long, or a total length for both of $8\frac{1}{2}''$. The total stock required for the forging would be:

Length stock for sides		....	$8\frac{1}{2}''$
" " "	end	....	$4\frac{11}{16}''$
Total " " "	forging	....	$13\frac{3}{16}''$.

Link.—As another example, take the link shown in Fig. 58. This may be divided into the two semicircles at the ends and the two straight sides. Calculating as always through the center of the stock, there are the two straight sides 2″ long, or 4″, and the two semicircular ends, or one complete circle for the two ends. The length required for these two ends would be $1\frac{1}{2}'' \times 3\frac{1}{7}'' = \frac{66}{14}'' = 4^{1}''$, or, nearly enough, $4\frac{11}{16}''$. The total length of the stock would be $4'' + 4\frac{11}{16}'' = 8\frac{11}{16}''$, to which must be added a slight amount for the weld.

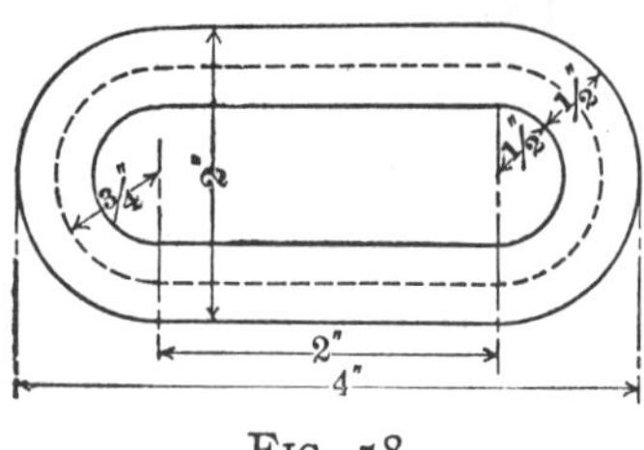

FIG. 58.

Double Link.—The double link in Fig. 59 is another example of stock calculation. Here there are two complete circles each having an inside diameter of $\frac{3}{4}''$, and, as they are made of $\frac{1}{4}''$ stock, a "calculating" diameter of 1″. The length of stock required for one side would be $3.1416'' \times 1'' = 3.1416''$, and the total length for complete links $3.1416'' \times 2'' = 6.2832''$, which is about $6\frac{1}{4}''$.

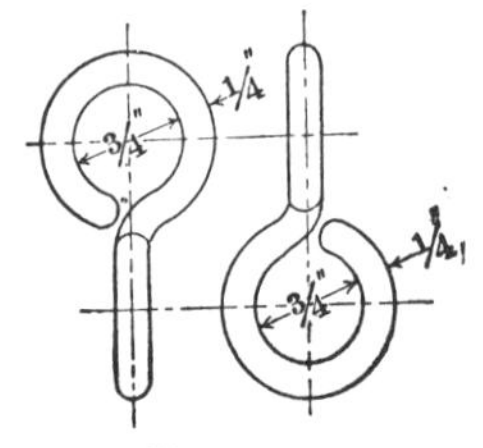

FIG. 59.

As a general rule it is much easier to make the calculations with decimals as above and then reduce these decimals to eighths, sixteenths, etc.

Use of Tables.—To aid in reduction a table of decimal equivalents is given on p. 249. By using this table it is only necessary to find the decimal result and select the nearest sixteenth in the table. It is generally sufficiently accurate to take the nearest sixteenth.

A table of circumferences of circles is also given; and by looking up the diameter of any circle the circumference may be found opposite.

To illustrate, suppose it is necessary to find the amount of stock required to make a ring 6″ inside diameter out of $\frac{3}{4}$″ round stock. This would make the calculating diameter of the ring $6\frac{3}{4}$.

In the table of circumferences and areas of circles opposite a diameter $6\frac{3}{4}$ is found the circumference 21.206. In the table of decimal equivalents it will be seen that $\frac{3}{16}$ is the nearest sixteenth to the decimal .206; thus the amount of stock required is $21\frac{3}{16}$″. This of course makes no allowance for welding.

Allowance for Welding.—Some allowance must always be made for welding, but the exact amount is very hard to determine, as it depends on how carefully the iron is heated and how many heats are taken to make the weld.

The only stock which is really lost in welding, and consequently the only waste which has to be allowed for, is the amount which is burned off or lost in scale when heating the iron.

Of course when preparing for the weld the ends of the piece are upset and the work consequently shortened, and the pieces are still farther shortened by overlapping the ends in making the weld; but all this material is afterward hammered back into shape so that no loss occurs here at all, except of course the loss from scaling.

A skilled workman requires a very small allowance for waste in welding, in fact sometimes none at all; but by the beginners an allowance should always be made.

No rules can be given; but as a rough guide on small work, a length of stock equal to from one-fourth to three-fourths the thickness of the bar will probably be about right for waste on rings, etc. When making straight welds, when possible it is better to allow a little more than is necessary and trim off the extra stock from the end of the finished piece.

Work of this kind should be watched very closely and the stock measured before and after welding in order to determine exactly how much stock is lost in welding. In this way an accurate knowledge is soon obtained of the proper allowance for waste.

CHAPTER IV.

UPSETTING, DRAWING OUT, AND BENDING.

Drawing Out.—When a piece of metal is worked out, either by pounding or otherwise, in such a way that the length is increased, and either the width or thickness reduced, we say that the metal is being "drawn out," and the operation is known as "drawing out."

It is always best when drawing out to heat the metal to as high a heat as it will stand without injury. Work can sometimes be drawn out much faster by working over the horn of the anvil than on the face, the reason being this: when a piece of iron is laid flat on the anvil face and hit a blow with the hammer, it flattens out and spreads both lengthwise and crosswise, making the piece longer and wider. The piece is not wanted wider, however, but only longer, so it is necessary to turn it on edge and strike it in this position, when it will again increase in length and also in thickness, and will have to be thinned out again. A good deal of work thus goes to either increasing the width or thickness, which is not wanted increased; consequently this work and the work required to again thin the forging are lost. In other words, when drawing out iron on the face of the anvil the force of the blow is expended in forcing the iron sidewise as well as lengthwise, and the work used in forcing the iron sidewise is lost. Thus only about one half the force of the blow is really used to do the work wanted.

Suppose the iron be placed on the horn of the

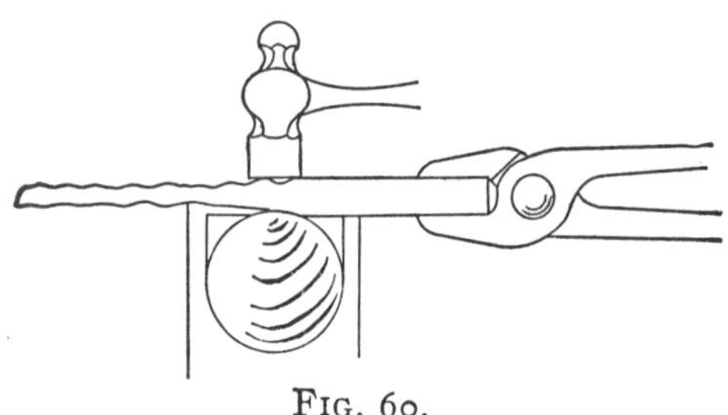

FIG. 60.

anvil, as shown in Fig. 60, and hit with the hammer as before. The iron will still spread out sidewise a little, but not nearly as much as before and will lengthen out very much more. The horn in this case acts as sort of a blunt wedge, forcing out the metal in the direction of the arrow, and the force of the blow is used almost entirely in lengthening the work.

Fullers may be used for the same purpose, and the work held either on the horn or the face of the anvil.

Drawing Out and Pointing Round Stock.—When drawing out or pointing round stock it should always be first forged down *square* to the required size and then, in as few blows as possible, rounded up.

Fig. 61 illustrates the different steps in drawing out round iron to a smaller size. *A* is the original bar, *B* is the first step, *C* is the next, when the iron is forged octagonal, and the last step is shown at *D*,

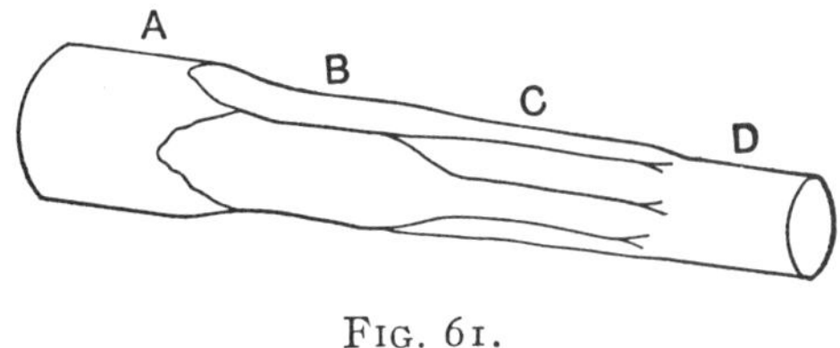

FIG. 61.

where the iron is finished up round. In drawing out a piece of round iron it should first be forged like *B*, then like *C*, and lastly finished like *D*.

As an example: Suppose part of a bar of $\frac{3}{4}''$ round stock is to be drawn down to $\frac{3}{8}''$ in diameter. Instead of pounding it down round and round until the $\frac{3}{8}''$ diameter is reached, the part to be drawn out should be forged perfectly square and this drawn down to $\frac{3}{8}''$, keeping it as nearly square as possible all the time.

The corners of the square are forged off, making an octagon, and, last of all, the work is rounded up. This prevents the metal from splitting, as it is very liable to do if worked round and round.

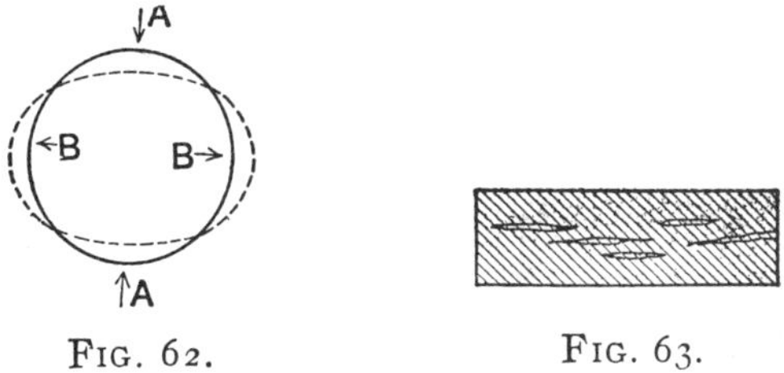

FIG. 62. FIG. 63.

The reason for the above is as follows: Suppose Fig. 62 represents the cross-section of a round bar as it is being hit on the upper side. The arrows indicate the flow of the metal—that is, it is forged together at *AA* and apart at *BB*. Now, as the bar is turned and the hammering continued, the outside metal is forced away from the center, which may, at last, give way and form a crack; and by the time the bar is of the required size, if cut, it would probably look something like Fig. 63.

The same precaution must be taken when forging any shaped stock down to a round or conical point. The point must first be made square and then rounded up by the method given above. If this is not done the point is almost *sure* to split.

Squaring Up Work.—A common difficulty met with in all drawing out, or in fact in all work which must be hammered up square, is the liability of the bar to forge into a diamond shape, or to have one corner projecting out too far. If a section be cut through a bar misshaped in this way, at right angles to its length, instead of being a square or rectangle, the shape will appear something like one of the outlines in Fig. 64.

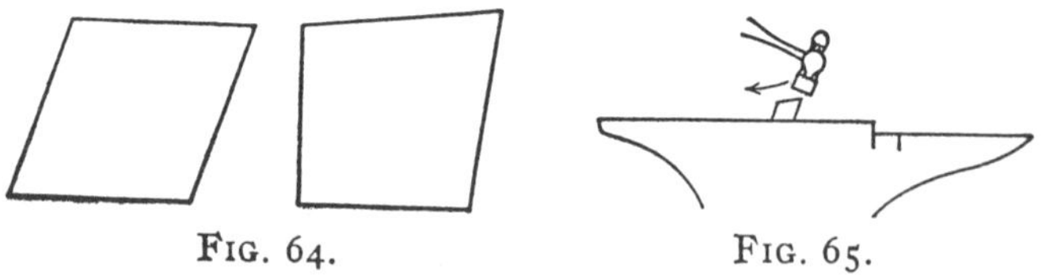

FIG. 64. FIG. 65.

To remedy this and square up the bad corners, lay the bar across the anvil and strike upon the projecting corners as shown in Fig. 65, striking in such a way as to force the extra metal back into the body of the bar, gradually squaring it off. Just as the hammer strikes the metal it should be given a sort of a sliding motion, as indicated by the arrow.

No attempt should be made to square up a corner of this kind by simply striking squarely down upon the work. The hammering should all be done in such a way as to force the metal back into the bar and away from the corner.

Upsetting.—When a piece of metal is worked in such a way that its length is shortened, and either or both its thickness and width increased, the piece is said to be upset; and the operation is known as upsetting.

There are several ways of upsetting, the method depending mostly on the shape the work is in. With short pieces the work is generally stood on end on the anvil and the blow struck directly on the upper end. The work should always be kept straight; after a few blows it will probably start to bend and must then be straightened before more upsetting is done.

If one part only of a piece is to be upset, then the heat must be confined to that part, as the part of the work which is hottest will be upset the most.

When upsetting a short piece for its entire length, it will sometimes work up like Fig. 66. This may be due to two causes: either the ends were hotter than the center or the blows of the hammer were too light. To bring a piece of this sort to uniform size throughout, it should be heated to a higher heat in the center and upset with heavy blows. If the work is very short it is not always convenient to confine the heat to the central part; in such a case, the piece may be heated all over, seized by the tongs in the middle and the ends cooled, one at a time, in the water-bucket.

FIG. 66. FIG. 67.

When light blows are used the effect of the blow does not reach the middle of the work, and consequently the upsetting is only done on the ends.

The effect of good heating and heavy blows is shown in Fig. 67. With a heavy blow the work is upset more in the middle and less on the ends.

To bring a piece of this kind to uniform size throughout, one end should be heated and upset and then the other end treated in the same way, confining the heat each time as much as possible to the ends.

Long work may be upset by laying it across the face of the anvil, letting the heated end extend two or three inches over the edge, the upsetting being done by striking against this end with the hammer or sledge. If the work is heavy the weight will offer enough resistance to the blow to prevent the piece from sliding back too far at each blow; but with lighter pieces it may be necessary to "back up" the work by holding a sledge against the unheated end.

Another way of upsetting the ends of a heavy piece is to "ram" the heated end against the side of the anvil by swinging the work back and forth horizontally and striking it against the side of the anvil. The weight of the piece in this case takes the place of the hammer and does the upsetting.

Heavy pieces are sometimes upset by lifting them up and dropping or driving them down on the face of the anvil or against a heavy block of iron resting on the floor. Heavy cast-iron plates are sometimes set in the floor for this purpose, and are called "upsetting-plates."

Fig. 68 shows the effect of the blows when upsetting the end of a bar. The lower piece has been properly heated and upset with heavy blows, while the other piece shows the effect of light blows. This last shape may also be caused by having the extreme end at a higher heat than the rest of the part to be upset.

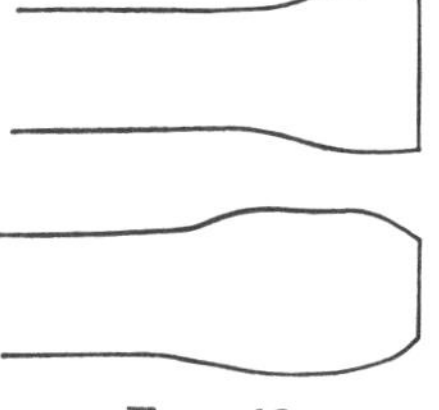

FIG. 68.

Punching.—There are two kinds of punches used for making holes in hot metal—the straight hand-punch, used with a hand-hammer, and the punch made from heavier stock and provided with a handle, used with a sledge-hammer.

Punches should, of course, be made of tool-steel.

For punching small holes in thin iron a hand-punch is ordinarily used. This is simply a bar of round or octagonal steel, eight or ten inches long, with the end forged down tapering, and the extreme end the same shape, but slightly smaller than the hole which is to be punched. Such a punch is shown in Fig. 69. The punch should taper uniformly, and the extreme end should be perfectly square across, not in the least rounding.

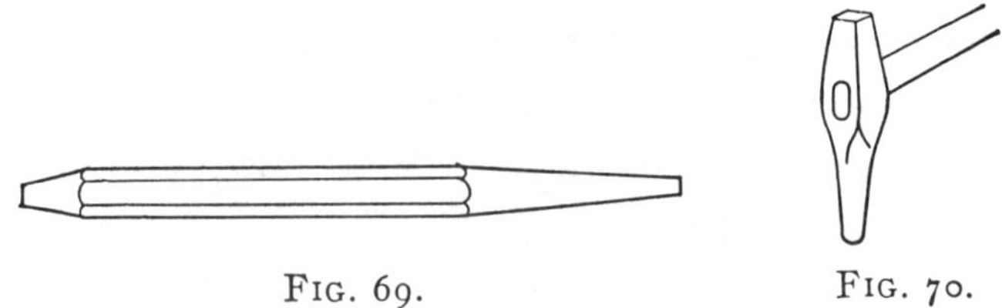

FIG. 69. FIG. 70.

For heavier and faster work with a helper, a punch like Fig. 70 is used. This is driven into the work with a sledge-hammer.

A, *B*, and *C*, in Fig. 71, show the different steps in punching a clean hole through a piece of hot iron.

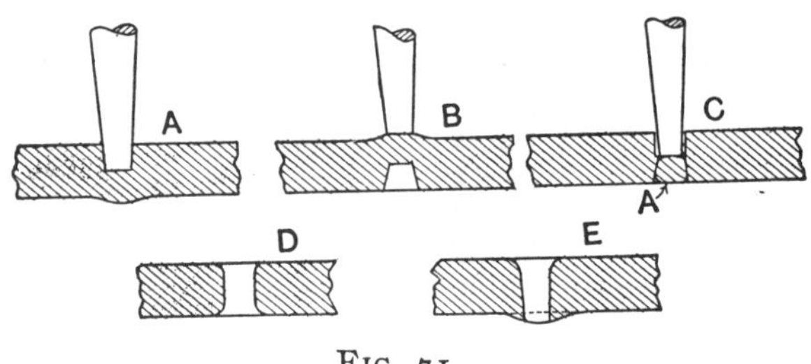

FIG. 71.

The punch is first driven about half-way through the bar while the work is lying flat on the anvil; this compresses the metal directly underneath the punch and raises a slight bulge on the opposite side of the bar by which the hole can be readily located. The piece is then turned over and the punching completed from this side, the small piece, "*A*", being driven completely through. This leaves a clean hole; while if the punching were all done from one side, a burr, or projection, would be raised on the side where the punch came through.

D and *E* (Fig. 71) illustrate the effects of proper and improper punching. If started from one side and finished from the other the hole will be clean and sharp on both sides of the work; but if the punching is done from one side only a burr will be raised, as shown at *E*, on the side opposite to that from which the punching is done.

If the piece is thick the punch should be started, then a little powdered coal put in the hole, and the punching continued. The coal prevents the punch from sticking as much as it would without it.

Bending.—Bends may be roughly divided into two classes—curves and angles.

Angles.—In bending angles it is nearly always necessary to make the bend at some definite point on the stock. As the measurements are much easier made while the stock is cold than when hot, it is best to "lay off" the stock before heating.

The point at which the bend is to be made should be marked with a center punch—generally on the edge of the stock, in preference to the side.

Marking with a cold chisel should not be done unless done very lightly on the *edge* of stock. If a slight nick be made on the side of a piece of stock to be bent, and the stockbent at this point with the nick outside, the small nick will expand and leave quite a crack. If the nick be on the inside, it is apt to start a bad cold shut which may extend nearly through the stock before the bending is finished.

Whenever convenient, it is generally easier to bend in a vise, as the piece may be gripped at the exact point where the bend is wanted.

When making a bend over the anvil the stock should be laid flat on the face, with the point at

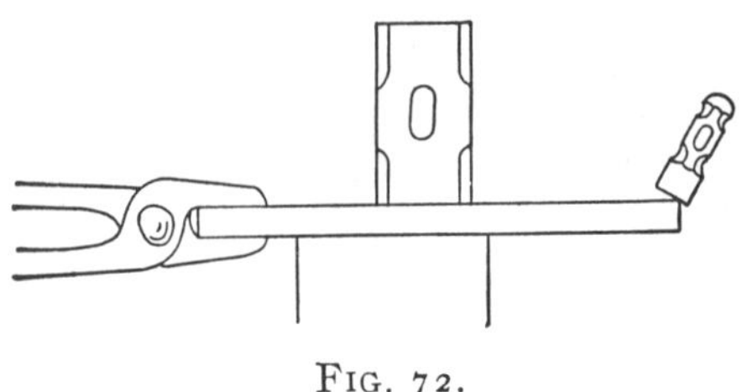

FIG. 72.

which the bend is wanted almost, but not quite, up to the outside edge of the anvil.

The bar should be held down firmly on the anvil by bearing down on it with a sledge, so placed that the outside edge of the sledge is about in line with the outside edge of the anvil.

This makes it possible to make a short bend with less hammering than when the sledge is not used.

The bar will pull over the edge of the anvil slightly when bending.

Bend with Forged Corner. — Brackets and other forgings are sometimes made with the outside corner of the bend forged up square, as shown in Fig. 73.

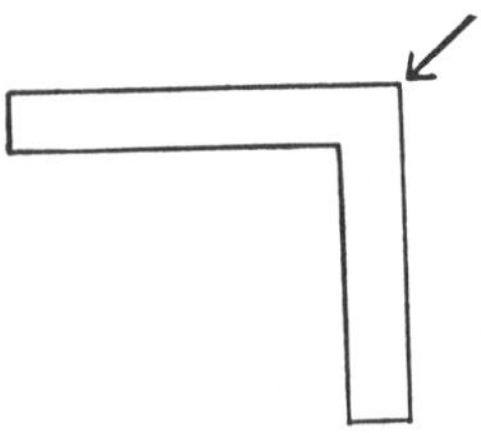

FIG. 73.

There are several ways of bending a piece to finish in this shape.

One way is to take stock of the required finished

size and bend the angle, forging the corner square as it is bent; another is to start with stock considerably thicker than the finished forging and draw down both ends to the required finished thickness, leaving a thin-pointed ridge across the bar at the point where the bend will come, this ridge forming the outside or square corner of the angle where the piece is bent; or this ridge may be formed by upsetting before bending.

The process in detail of the first method mentioned is as follows: The first step is to bend the bar so that it forms nearly a right angle, keeping the bend as sharp as possible, as shown at *A* (Fig. 74).

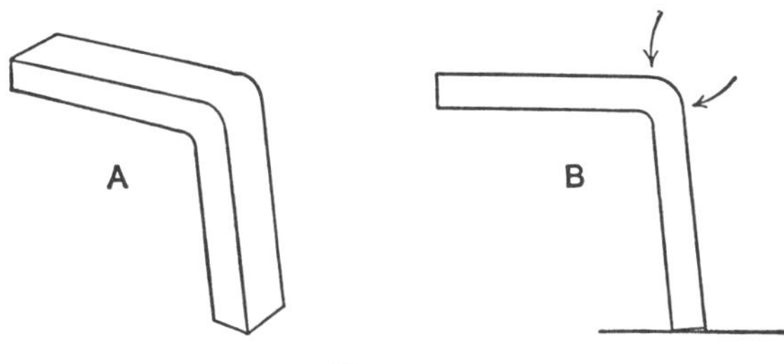

FIG. 74.

This should be done at a high heat, as the higher the heat the easier it is to bend the iron and consequently the sharper the bend.

Working the iron at a good high heat, as before, the outside of the bend should be forged into a sharp corner, letting the blows come in such a way as to force the metal out where it is wanted, being careful not to let the angle bend so that it becomes less than a right angle or even equal to one. Fig.

74, *B*, shows the proper way to strike. The arrows indicate the direction of the blows.

The work should rest on top of the anvil while this is being done, not over one corner. If worked over the corner, the stock will be hammered too thin.

The object in keeping the angle obtuse is this: The metal at the corner of the bend is really being upset, and the action is somewhat as follows: In Fig. 75 is shown the bent piece on the anvil. We will suppose the blows come on the part *A* in the direction indicated by the heavy arrow. The metal, being heated to a high soft heat at *C*, upsets, part of it forming the sharp outside corner and part flowing as shown by the small arrow at *C* and

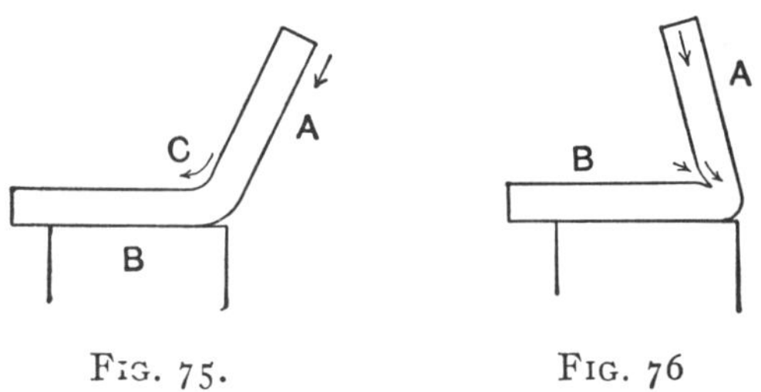

FIG. 75. FIG. 76

making a sort of fillet on the inside corner. If in place of having the angle greater than 90 degrees it had been an acute angle (Fig. 76), the metal forced downward by the blows on *A* would carry with it part of the metal on the inside of the piece *B*, and a cold shut or crack would be formed on the inside of the angle. To form a sound bend the corner must be forged at an angle greater than a right angle. When the piece has been brought to a sharp corner the last step is to square up the bend over the corner, or edge, of the anvil.

The second way of making the above is to forge a piece as shown in Fig. 77, where the dotted lines indicate the size of the original piece. This piece is then bent in such a way that the ridge, *C*, forms the outside sharp corner of the angle.

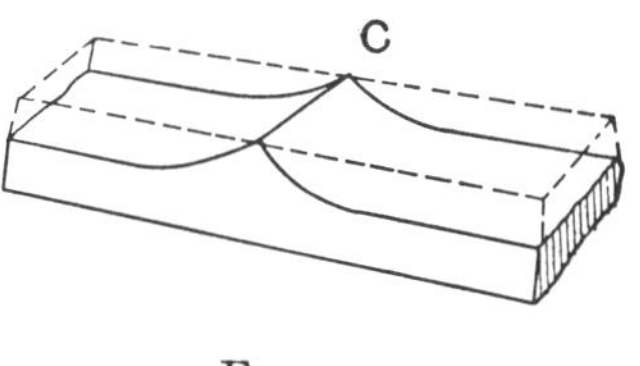

FIG. 77.

This ridge is sometimes upset in place of being drawn out.

The first method described is the most satisfactory.

Ring-bending.—In making a ring the first step of course is to calculate and cut from the bar the proper amount of stock. The bend should always be started from the end of the piece. For ordinary rings up to 4″ or 5″ in diameter the stock should be heated for about one-half its length. To start bending, the extreme end of the piece should be first bent over the horn of the anvil, and the bar should be fed across the horn of the anvil and bent down as it is pushed forward. Do not strike directly on top of the horn, but let the blows fall a little way from it, as in Fig. 78. This bends the iron and does not pound it out of shape. One-half of the ring is bent in this way and then the part left straight is heated. This half is bent up the same as the other, starting from the end exactly as before.

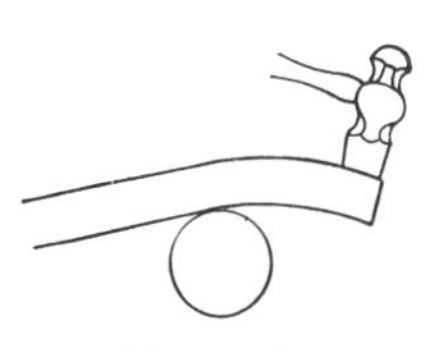

FIG. 78.

Eye-bending. — The first step in making an eye like Fig. 79 is to calculate the amount of stock required for the bend. The amount required in this case, found by looking up the circumference of a 2″ circle in the table, is $7\frac{1}{2}''$. This distance should be laid off by making a chalk-mark on the face of the anvil $7\frac{1}{2}''$ from the left-hand end.

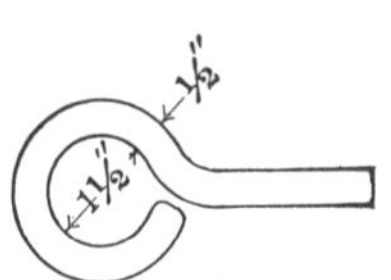

Fig. 79.

A piece of iron is heated and laid on the anvil with the heated end on the chalk-mark, the rest of the bar extending to the left. A hand-hammer is held on the bar with the edge of the hammer directly in line with the end of the anvil. This measures off $7\frac{1}{2}''$ from the edge of the hammer to the end of the bar. The bar is then laid across the anvil bringing the edge of the hammer exactly in line with the outside edge of the anvil, thus leaving $7\frac{1}{2}''$ projecting over the edge. This projecting end is bent down until it forms a right angle. The extreme end of this bent part is then bent over the horn into

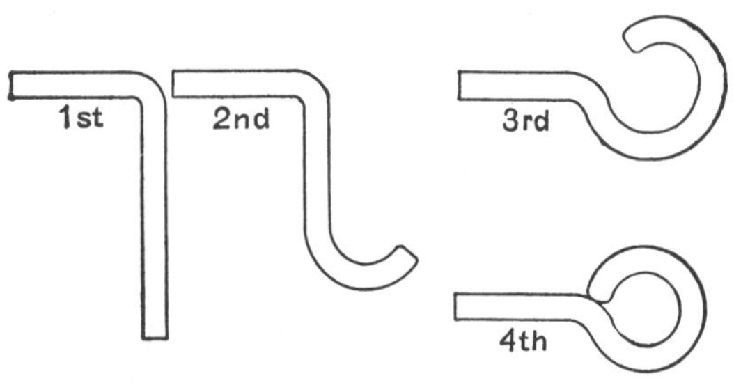

Fig. 80.

the circular shape and the bending continued until the eye is formed.

The same general method as described for bending rings should be followed. The different steps are shown in Fig. 80.

If an eye is too small to close up around the horn, it may be closed as far as possible in this way, and then completely closed over the corner or on the face of the anvil, as shown in Fig. 81.

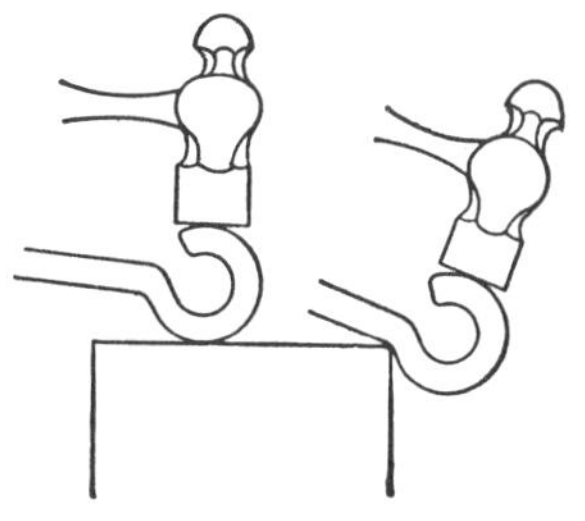

FIG. 81.

Double Link.—Another good example of this sort of bending is the double link, shown in Fig. 59.

The link is started by bending the stock in the *exact center*, the first step being to bend a right angle. This step, with the succeeding ones, is shown in Fig. 82.

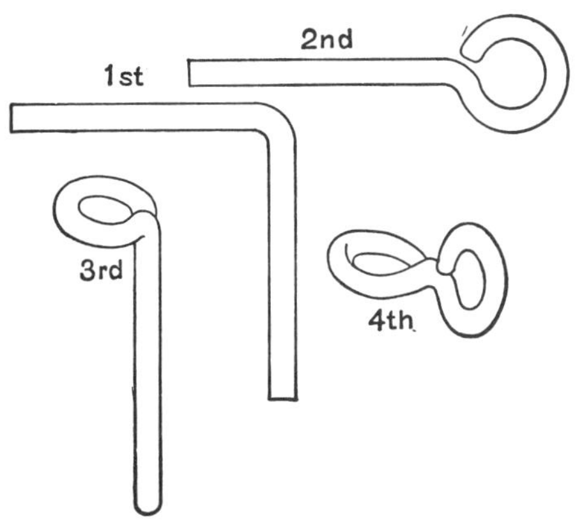

FIG. 82.

After this piece has been bent into a right angle, the ring on the end should be bent in the same way as an ordinary ring; excepting that all the bending is done from one end of the piece, starting from the extreme end as usual.

Twisting.—Fig. 83 shows the effects produced by twisting stock of various shapes—square, octagonal,

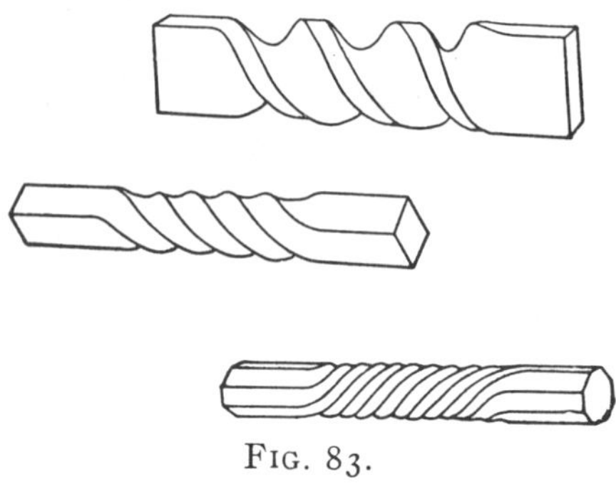

FIG. 83.

and flat, the shapes being shown by the cuts in each case.

To twist work in this way it should be brought to a uniform heat through the length intended to twist. When the bar is properly heated it should be firmly gripped with a pair of tongs, or in a vise, at the exact point where the twist is to commence. With another pair of tongs the work is taken hold of where the twist is to stop, and the bar twisted through as many turns as required. The metal will of course be twisted only between the two pairs of tongs, or between the vise and the tongs, as the case may be; so care must be used in taking hold of the bar or the twist will be made at the wrong points.

The heat must be the same throughout the part to be twisted. If one part is hotter than another, this hotter part, being softer, will twist more easily, and the twist will not be uniform. If one end of the bar is wanted more tightly twisted than the other, the heat should be so regulated that the part is heated hottest that is wanted tightest twisted; the heat gradually shading off into the parts wanted more loosely twisted.

Reverse Twisting. — The effect shown in Fig. 84 is produced by reversing the direction of twisting.

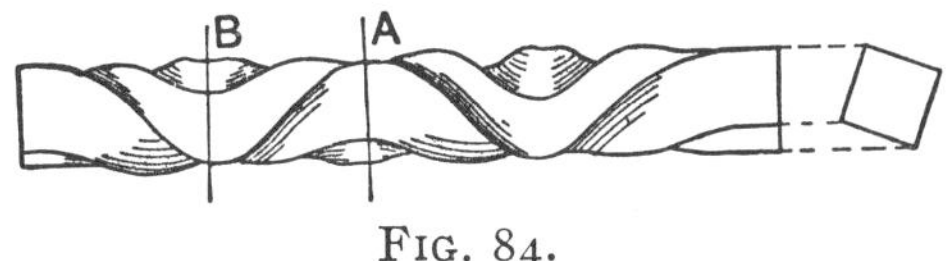

FIG. 84.

A square bar is heated and twisted enough to give the desired angle. It is then cooled, in as sharp a line as possible, as far as *B*, and twisted back in the opposite direction. It is again heated, cooled up to *A*, and twisted in the first direction; and this operation is continued until the twist is of the desired length.

CHAPTER V.

SIMPLE FORGED WORK.

Twisted Gate-hook.—This description answers, of course, not only for this particular piece, but for others of a like nature.

Fig. 85 shows the hook to be made. To start with, it must be determined what length of stock, after it is forged to proper size, will be required to bend up the ends.

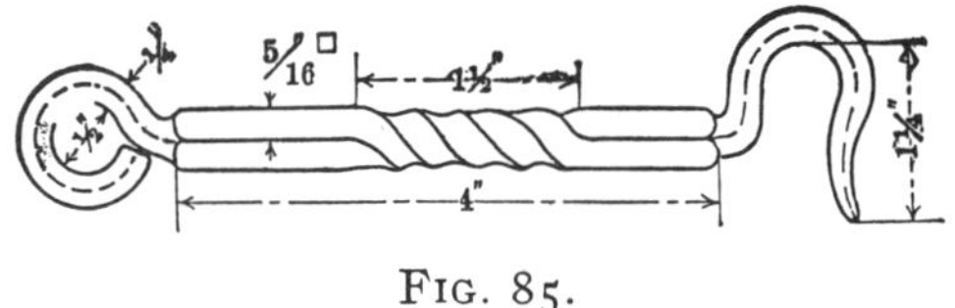

FIG. 85.

The length of straight stock necessary should, of course, be measured through the center of the stock on the dotted lines in the figure. To do this lay out the work full size, and lay a string or thin piece of soft wire upon the lines to be measured. It is then a very easy matter to straighten out the wire or string, and measure the exact length required. If the drawing is not made full size, an accurate sketch may be made on a board, or other flat surface, and the length measured from this.

The hook as above will require about $2\frac{7}{8}''$ length for stock; the eye, about $2\frac{5}{8}''$.

The first step would be like Fig. 86.

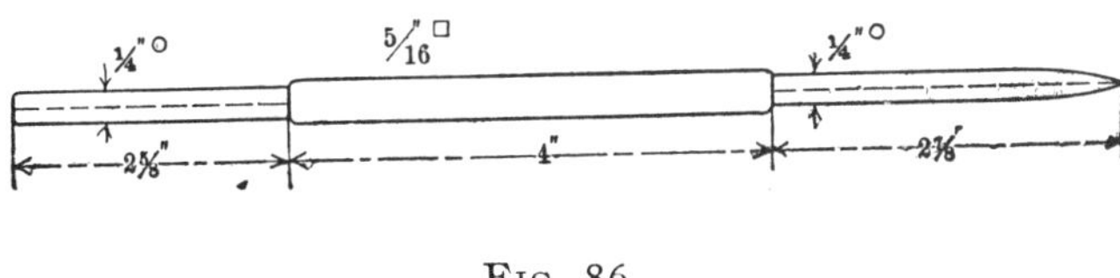

FIG. 86.

After cutting the piece of $\frac{5}{16}''$ square stock, start the forging by drawing out the end, starting from the end and working back into the stock until a piece is forged out $2\frac{5}{8}''$ long and $\frac{1}{4}''$ in diameter. Now work in the shoulder with the set-hammer in the following way:

Forming Shoulders: Both Sides—One Side.—Place the piece on the anvil in such a position that the point where the shoulder is wanted comes exactly on the nearest edge of the anvil. Place the set-hammer on top of the piece in such a way that its edge comes directly in line with the edge of the anvil (Fig. 87). Do *not* place the piece like

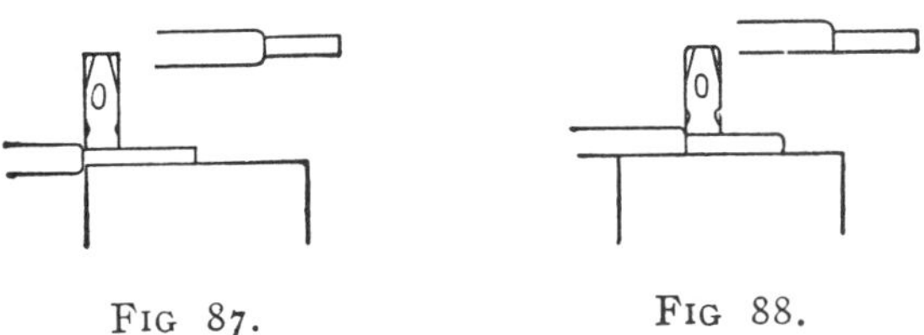

FIG 87. FIG 88.

Fig. 88, or the result will be as shown—a shoulder on one side only. As the shoulder is worked in the piece should be turned continually, or the shoulder will work in faster on one side than on the other Always be careful to keep the shoulder exactly even with the edge of the anvil.

When the piece is formed in the proper shape on one end, start the second shoulder 4″ from the first, and finish like Fig. 86. Bend the eye and then the hook; and, *lastly*, put the twist in the center. Make the twist as follows:

First make a chalk-mark on the jaws of the vise, so that when the end of the hook is even with the mark the edge of the vise will be where one end of the twist should come. Heat the part to be twisted to an even yellow heat (be sure that it is heated evenly); place it in a vise quickly, with the end even with the mark; grasp the piece with the tongs, leaving the distance between the tongs and vise equal to the length of twist (Fig. 89); and twist it around one complete turn.

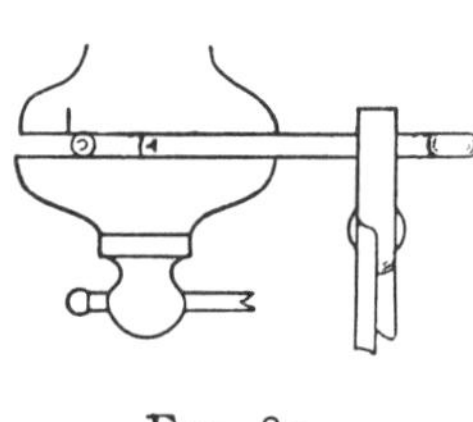

FIG. 89.

The eye should be bent as described before, and the hook bent in the same general way as the eye.

Grab-hooks.—This is the name given to a kind of hooks used on chains, and made for grabbing or hooking over the chain. The hook is so shaped that the throat, or opening, is large enough to slip easily over a link turned edgewise, but too narrow to slip down off this link on to the next one, which, of course, passes through the first link at right angles to it.

Grab-hooks are made in a variety of ways, one of which is given below in detail.

Fig. 90 will serve as an example. To forge this, use a bar of round iron large enough in section to form the heavy part of the hook. This bar should first be slightly upset, either by ramming or hammering, for a short distance from the end, and then flattened out like Fig. 91.

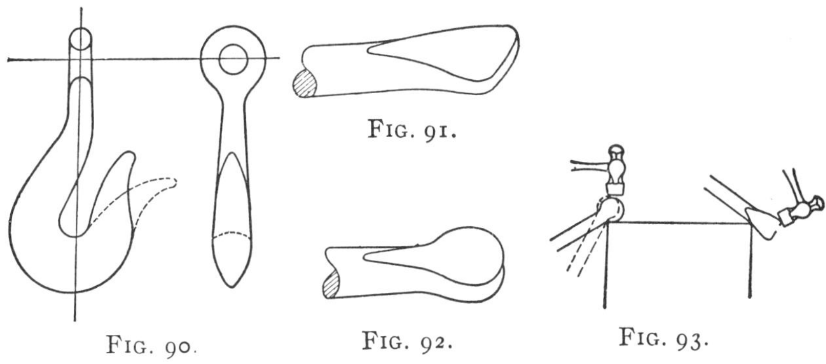

FIG. 90. FIG. 91. FIG. 92. FIG. 93.

The next step is to round up the part for the eye, as shown in Fig. 92, by forming it over the corner of the anvil as indicated in Fig. 93. The eye should be forged as nearly round as possible, and then punched.

Particular attention should be paid to this point. If the eye is not properly rounded before punching, it is difficult to correct the shape afterward.

After punching, the inside corners of the hole are rounded off over the horn of the anvil in the manner shown in Fig. 94. Fig. 95 shows the appearance of a section of the eye as left by the punch. When the eye is finished it should appear as though bent up from round iron—that is, all the square corners should be rounded off as shown in Fig. 96.

When the eye is completed the body of the hook

should be drawn out straight, forged to size, and then bent into shape. Care should be taken to

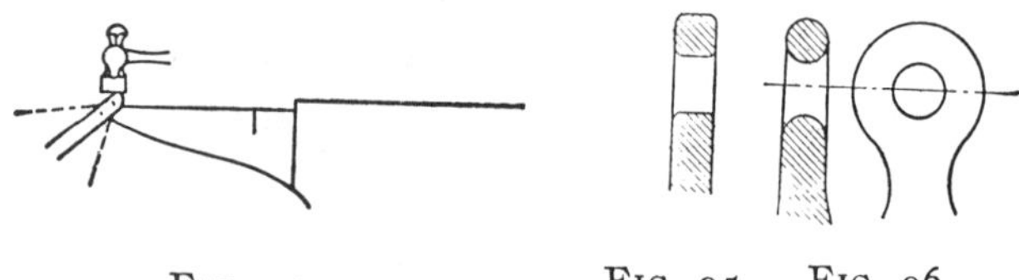

FIG. 94. FIG. 95. FIG. 96.

keep the hook thickest around the bottom of the bend.

As the stock is entirely formed before bending, the length of the straight piece must be carefully measured, as indicated at *A* (Fig. 97), where the piece is shown ready for bending. To determine the required length the drawing or sample should be measured with a string or piece of flexible wire, measuring along the center of the stock, from the extreme point to the center of the eye.

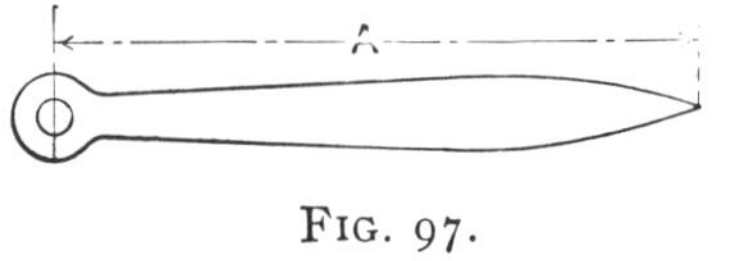

FIG. 97.

The weakest point of almost any hook is in the bottom bend. When the hook is strained there is a tendency for it to straighten out and take the shape shown by the dotted lines in Fig. 90. To avoid this the bottom of the hook must be kept as thick as possible along the line of strain, which is shown by the line drawn through the eye. A good shape for this lower bend is shown in the sketch, where it will be noticed that the bar has been hammered a little thinner in order to increase the thickness of the metal in the direction of the line of strain.

The part of the hook most liable to bend under a load is the part lying between the points marked *I* and *J* in Fig. 102.

Another style of grab-hook is shown in Fig. 98,

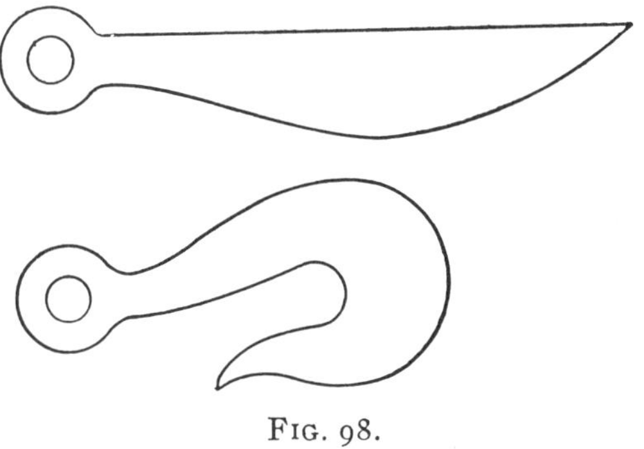

FIG. 98.

which shows the finished hook and also the straight piece ready for bending.

The forming will need no particular description. The hook shown is forged about $\frac{3}{8}''$ thick; the outside edge around the curve being thinned out to about $\frac{1}{8}''$, in order to give greater stiffness in the direction of the strain.

Stock about $\frac{3}{8}'' \times 1''$ is used.

A very convenient way to start the eye for a hook of this kind, or in fact almost any forged eye, is shown in Fig. 99. Two fullers, top and bottom, are used, and the work shaped as shown. The bar should be turned, edge for edge, between every few blows, if the grooves are wanted of the same depth. After cutting the grooves the edge is shaped the same as described above.

A grab-hook, sometimes used on logging-chains, is shown in Fig. 100. This is forged from square stock by flattening and forming one end into an eye and pointing the other end; after which the hook is bent into shape.

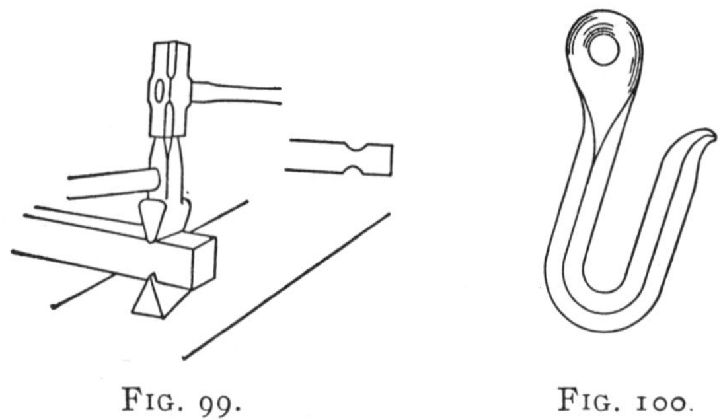

FIG. 99. FIG. 100.

Welded Eye-hooks. — Hooks sometimes have the eye made by welding instead of forging from the solid stock. Such a hook is shown in Fig. 101,

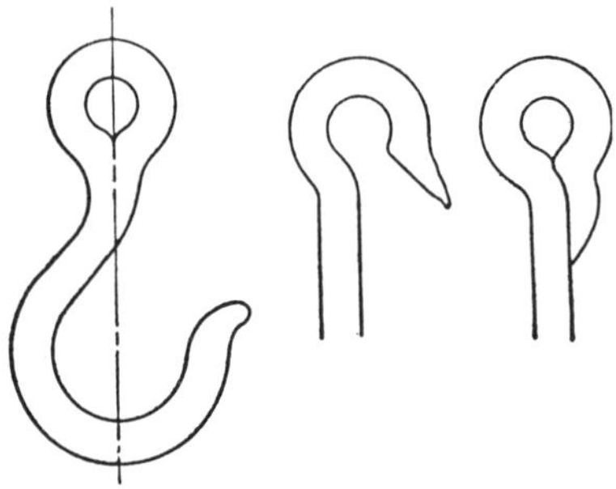

FIG. 101.

which also shows the stock scarfed and bent into shape ready for closing up the eye for the weld, and also the eye ready for welding. Before heating for the weld, the eye should be closed, and stock at the end be bent close together. The scarf should be pointed the same as for any other round weld.

This sort of eye is not as strong as a forged eye of the same size; but is usually as strong as the rest of the hook, as the eye is generally considerably stronger than any other part.

Hoisting-hooks. — A widely accepted shape for hoisting-hooks, used on cranes, etc., is shown in Fig. 102. The shape and formula are given by Henry R. Town, in his Treatise on Cranes.

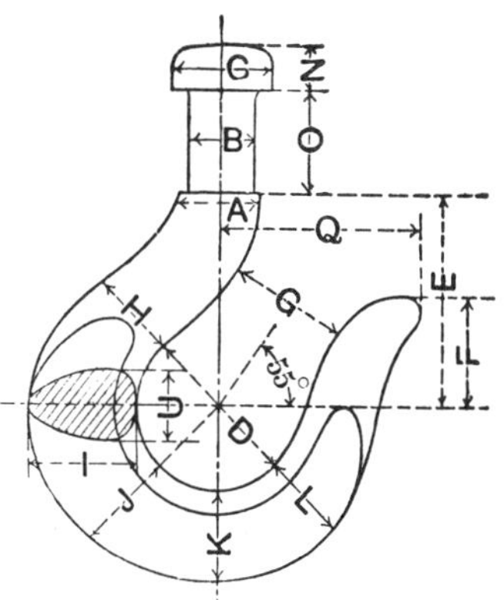

FIG. 102.

T = working load in tons of 2000 lbs.

A = diameter of round stock used to form hook.

The size of stock to use for a hook to carry any particular load is given below. The capacity of the hook, in tons, is given in the upper line—the figures in the lower line, directly under any particular load in the upper line, giving the size of bar required to form a hook to be used at that load.

$T=\frac{1}{8}$	$\frac{1}{4}$	$\frac{1}{2}$	1	$1\frac{1}{2}$	2	3	4	5	6	8	10
$A=\frac{5}{8}$	$\frac{11}{16}$	$\frac{3}{4}$	$1\frac{1}{16}$	$1\frac{1}{4}$	$1\frac{3}{8}$	$1\frac{3}{4}$	2	$2\frac{1}{4}$	$2\frac{1}{2}$	$2\frac{7}{8}$	$3\frac{1}{4}$

The other dimensions of the hook are found by the following formula, all the dimensions being in inches:

$$D = .5T + 1.25$$
$$E = .64T + 1.6$$
$$F = .33T + .85$$

$$
\begin{aligned}
G &= .75D \\
O &= .363T + .66 \\
Q &= .64T + 1.6 \\
H &= 1.08A \\
I &= 1.33A \\
J &= 1.2A \\
K &= 1.13A \\
L &= 1.05A \\
M &= .5A \\
N &= .85B - .16 \\
U &= .866A
\end{aligned}
$$

To illustrate the use of the table, suppose a hook is wanted to raise a load of 500 lbs.

In the line marked T in the table are found the figures $\frac{1}{4}$, denoting a load of one-quarter of a ton, or 500 lbs. Under this are the figures $\frac{11}{16}$, giving the size stock required to shape the hook.

The different dimensions of the hook would be found as follows:

$$D = .5 \times {}^{1}/_{4} + 1\,{}^{1}/_{4}'' = 1\,{}^{3}/_{8}''.$$
$$E = .64 \times {}^{1}/_{4} + 1.6'' = 1.76 = 1\,{}^{3}/_{4}'' \text{ about.}$$

. . . .

$$H = 1.08A = 1.08 \times {}^{11}/_{16} = .74 = {}^{3}/_{4} \text{ about.}$$
$$I = 1.33A = 1.33 \times {}^{11}/_{16} = .915 = {}^{29}/_{32} \text{ about.}$$

When reducing the decimals, the dimensions which have to do only with the bending of the hook, that is, the opening, the length, the length of point, etc., may be taken as the nearest 16th, but these dimensions for flattening should be reduced to the nearest 32d on small hooks.

The complete dimensions for the hook in question, 1000 lbs. capacity, would be as follows:

$D = 1\,^3/_8''$	$G = 1''$	$H = ^3/_4''$	$L = ^{23}/_{32}''$
$E = 1\,^3/_4''$	$O = ^3/_4''$	$I = ^{29}/_{32}''$	$M = ^{11}/_{32}''$
$F = ^{15}/_{16}''$	$Q = 1\,^3/_4''$	$J = ^{13}/_{16}''$	
		$K = ^{25}/_{32}''$	$U = ^9/_{16}''$

Bolts.—Bolts are made by two methods, upsetting and welding. The first method is the more common, particularly on small bolts, where it is nearly always used, the stock being upset to form the head. In the second method the head is formed by welding a ring of stock around the stem.

An upset head is stronger than a welded head, provided they are both equally well made.

The size of the bolt is always given as the diameter and length of the shank, or stem. Thus, a $\frac{1}{2}''$ bolt, $6''$ long, means a bolt having a shank $\frac{1}{2}''$ in diameter, and $6''$ long from the under side of the head to the end.

Dimensions of bolt-heads are determined from the diameter of the shank, and should always be the same size for the same diameter, being independent of the length.

The diameter and thickness of the head are measured as shown in Fig. 103.

The dimensions of both square and hexagonal heads are as follows:

D = diameter of head across the flats (short diameter).

T = thickness of head.

S = diameter of shank of bolt.

$D = 1\frac{1}{2} \times S + \frac{1}{8}''.$

$T = S.$

For a 2″ bolt the dimensions would be calculated as follows:

Diameter of head would equal $1\frac{1}{2}'' \times 2'' + \frac{1}{8}'' = 3^1/_8''$.

Thickness of head would be 2″.

These are dimensions for rough or unfinished heads; each dimension of a finished head is $^1/_{16}''$ less than the same dimension of the rough head.

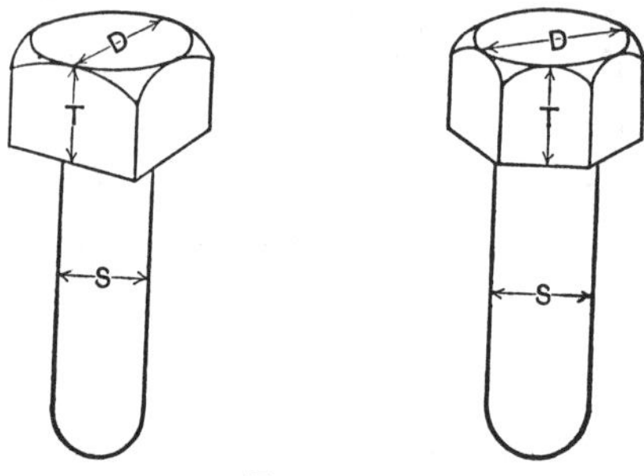

FIG. 103.

Bolts generally have the top corners of the head rounded or chamfered off (Fig. 103). This can be done with a hand-hammer, or with a cupping-

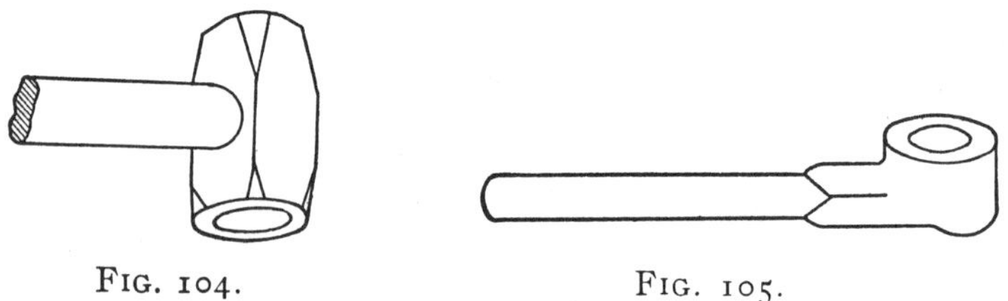

FIG. 104. FIG. 105.

tool (Fig. 104), which is simply a set-hammer with the bottom face hollowed out into a bowl or cup shape.

For making bolts one special tool is required, the heading-tool. This is commonly made something the shape of Fig. 105, although for a "hurry-up" bolt sometimes any flat strip of iron with a hole punched the proper size to admit the stem of the bolt can be used.

When in use this tool is placed on the anvil directly over the square hardie-hole, the stem of the bolt projecting down through the heading-tool and hardie-hole while the head is being forged on the bolt.

This heading-tool is made with one side of the head flush with the handle, the other side projecting a quarter of an inch or so above it. The tool should always be used with the *flat side on the anvil.*

Upset-head Bolt.—An upset head is made as follows: The stock is first heated to a high heat for a short distance at the end, and upset as shown at Fig. 106. The bolt is then dropped through the heading tool, the upset portion projecting above. This upset part is then flattened down on the tool as shown at *B*, and forged square or hexagonal on the anvil.

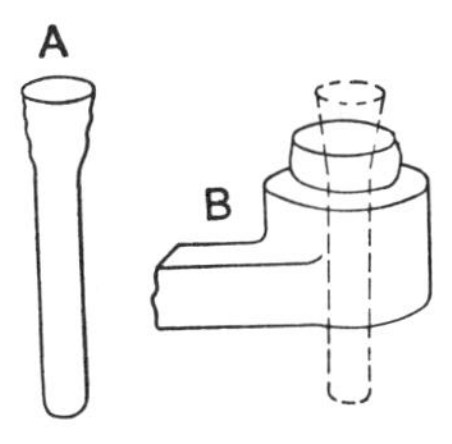

FIG. 106.

The hole in the heading-tool should be large enough to allow the stock to slip through it nearly up to the upset portion.

Welded-head Bolts.—A welded-head bolt is made by welding a ring of square iron around the shank to form the head, which is then shaped in a heading-tool the same as an upset head. A piece of square iron of the proper size is bent into a ring, but not welded. About the easiest way to do this is to take a bar several feet long, bend the ring on the end, and then cut it off as shown in Fig. 107.

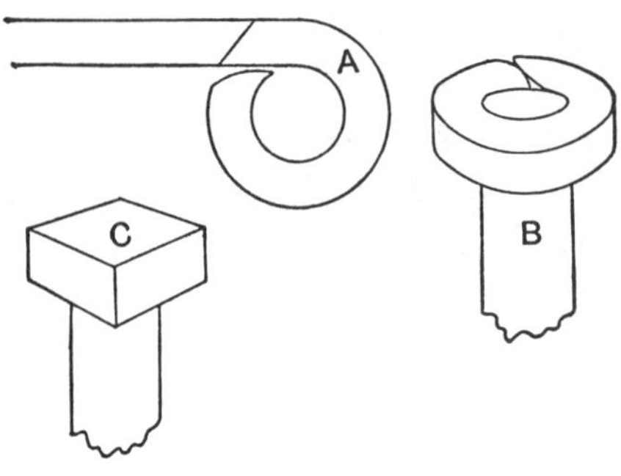

FIG. 107.

This ring is just large enough, when the ends are slightly separated, to slip easily over the shank.

The shank is heated to about a welding heat, the ring being slightly cooler, and the two put together as shown in Fig. 107, *B*. The head is heated and welded, and then shaped as described above.

When welding on the head it should be hammered *square* the first thing, and *not* pounded round and round. It is much easier to make a sound weld by forging square.

Care must be used when taking the welding heat to heat slowly, otherwise the outside of the ring will be burned before the shank is hot enough to stick.

It is sometimes necessary when heating the bolt-head for welding to cool the outside ring to prevent its burning before the shank has been heated sufficiently to weld; to do this put the bolt in the water sideways just far enough to cool the outside edge of the ring and leave the central part, or shank, hot.

Tongs.—Tongs are made in a great variety of ways, several of which are given below.

Common flat-jaw tongs, such as are used for holding stock up to about $\frac{3}{4}$ inch thick, may be made as follows: Stock about $\frac{3}{4}$ inch square should be used. This is first bent like *A*, Fig. 108. To form the eye the bent stock is laid

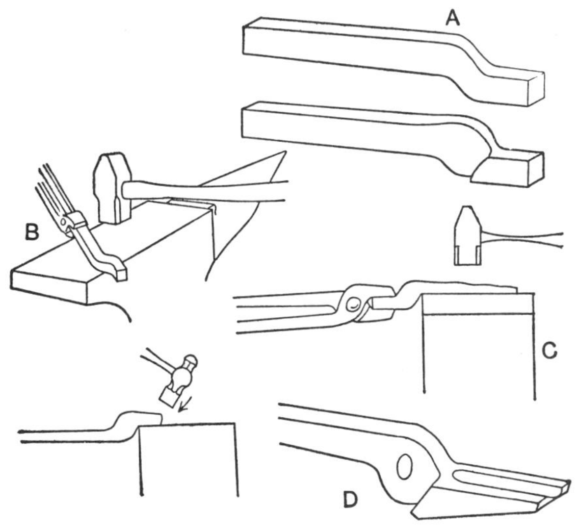

FIG. 108.

across the anvil in the position shown at *B*, and flattened by striking with a sledge the edge of the anvil, forming the shoulder for the jaw. A set-hammer may be used to do this by placing the piece with the other side up, flat on the face of the anvil, and holding the set-hammer in such a way as to form the shoulder with the edge of the hammer, the face of the hammer flattening the eye.

The long handle is drawn out with a sledge, working as shown at *C*. When drawing out work this way the forging should always be held with the straight side up, the corner of the anvil forming the sharp corner up against the shoulder on the piece. If the piece be turned the other side up, there is danger of striking the projecting shoulder with the sledge and knocking the work out of shape.

For finishing up into the shoulder a set-hammer or swage should be used, and the handles should be smoothed off with a flatter, or between top and bottom swages. The jaw may be flattened as shown at *D*.

The inside face of the jaw should be slightly creased with a fuller, as this insures the tongs gripping the work firmly with the sides of the jaws, and not simply touching it at one point in the center, as they sometimes do if this crease is not made.

After the tongs have been shaped, and are finished in every other way, the hole for the rivet should be punched. The rivet should drop easily into the hole. The straight end of the rivet should be brought to a high heat, the two parts of the tongs placed together with the holes in line, the rivet inserted, and the end "headed up." Most of the heading should be done with the pene end of the hammer. After riveting the tongs will probably be rather "stiff"; opening and shutting them several times, while the rivet is still red-hot, will leave them loose. The tongs should be finished by fitting to a piece of stock of the size on which they are to be used.

Light Tongs.—Tongs may be made from flat stock in the following way: A cut is made with a narrow fuller at the right distance from the end of the bar to leave enough stock to form the jaw between the cut and the end, as shown at *A*, Fig. 109.

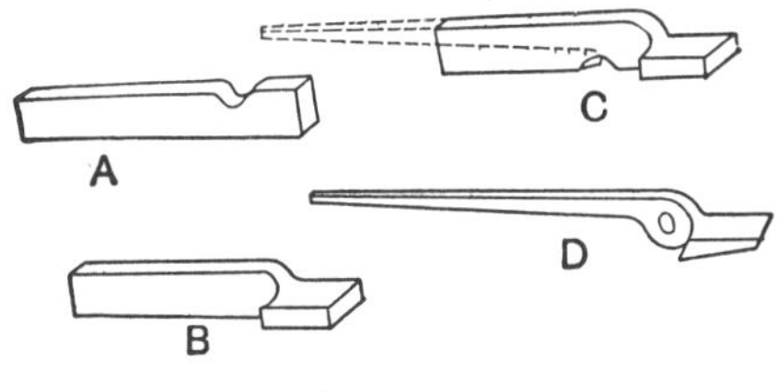

FIG. 109.

This end is bent over as shown at *B* and a second fuller cut made, shown at *C*, to form the eye. The other end of the bar is drawn out to form the handle, as indicated by the dotted lines. The jaw is shaped, the rivet-hole punched, and the tongs finished, as at *D*, in the usual way.

Tongs of this character may be used on light work.

Tongs for Round Stock. — Tongs for holding round stock may be made by either of the above methods, the operations in making being the same, with the exception of shaping the jaws, which may be done in this way: A top fuller and bottom swage are used, the swage being of the size to which it is wished to finish the outside of the jaws, the fuller

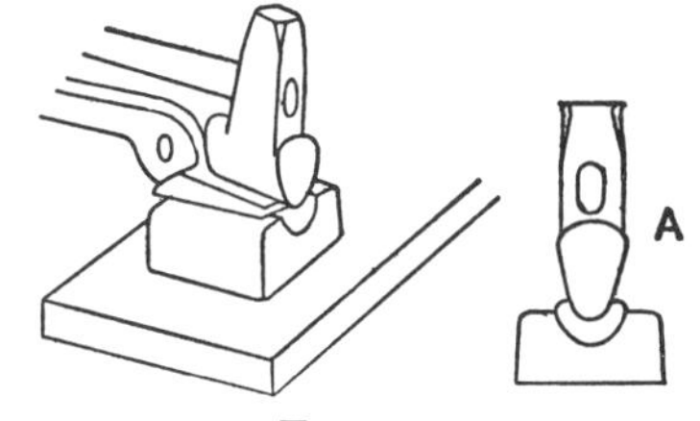

FIG. 110.

the size of the inside. The jaw is held on the swage, and the fuller placed on top and driven down on it, Fig. 110, forcing the jaw to take the desired shape, shown at *A*. The final fitting is done as usual, after the jaws are riveted together.

Welded Tongs.—Tongs with welded handles are made in exactly the same way as those with solid, drawn-out handles excepting that, in place of drawing out the entire length of the handle, a short stub only is forged, a few inches long, and to this is welded a bar of round stock to form the handle. Fig. 111 shows one ready for welding.

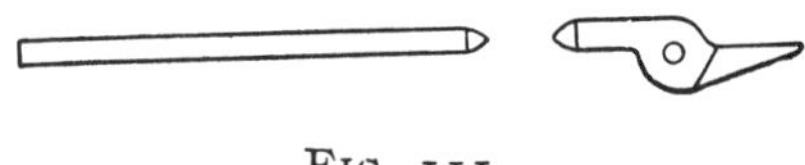

FIG. 111.

Pick-up Tongs. — No particular description is necessary for making pick-up tongs. The tongs may be drawn out of a flat piece and bent as shown in Fig. 112.

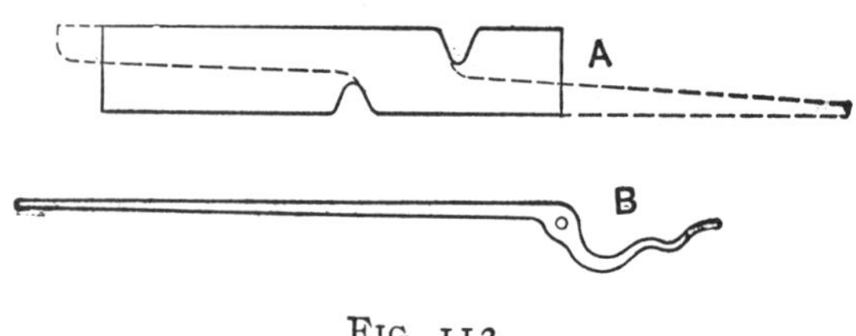

FIG. 112.

Bolt-tongs. — Bolt-tongs are easily made from round stock, although square may be used to advantage.

The first operation is to bend the bar in the shape

shown in Fig. 113. This may be done with a fuller over the edge of the anvil, as shown at *A*. When bending the extreme end of the jaw the bar should be held almost level at first, and gradually swung down, as shown by the arrow, until the end is properly bent.

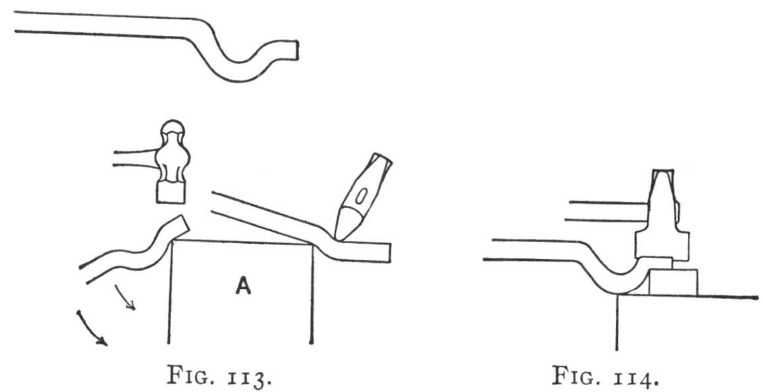

FIG. 113. FIG. 114.

The eye may be flattened with the set-hammer, and the part between the jaw proper and the eye worked down to shape over the horn and on the anvil with the same tool.

The jaw proper is rounded and finished with a fuller and swage, as shown in Fig. 114.

There is generally a tendency for the spring of the jaw to open up too much in forging. This may be bent back into shape either on the face of the anvil, as shown at *A* (Fig. 115), or over the horn, as at *B*.

Another method of making the first bend, when starting the tongs, is shown in Fig. 116. A swage-block and fuller are here used; a swage of the proper size could of course be used in place of the block.

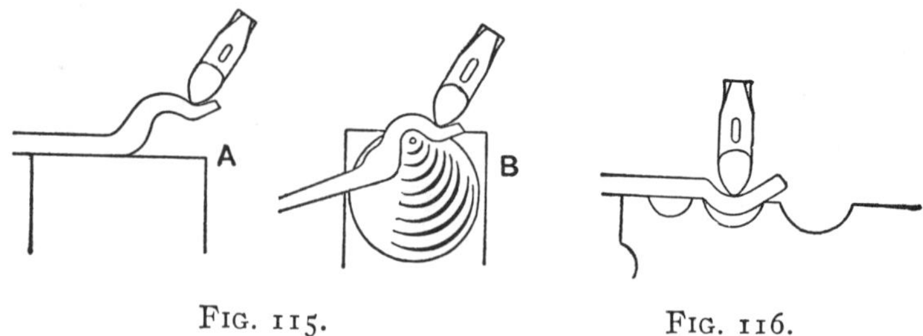

FIG. 115. FIG. 116.

Ladle.—A ladle, similar to Fig. 117, may be made of two pieces welded together, one piece forming the handle, the other the bowl.

A square piece of stock of the proper thickness is cut and "laid out" (or marked out) like Fig. 118; the center of the piece being first found by drawing the diagonals.

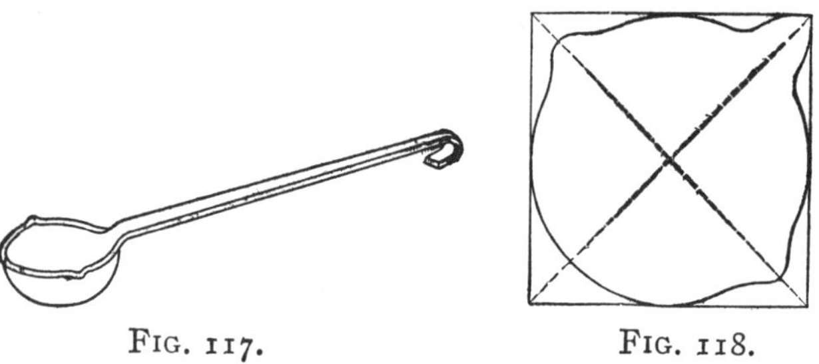

FIG. 117. FIG. 118.

A circle is drawn as large as possible, with its center on the intersection of the diagonals; the piece is cut out with a cold chisel to the circle, excepting at the points where projections are left for lips and for a place to weld on the handle. This latter projection is scarfed and welded to the strip forming the handle.

The bowl is formed from the circular part by heating it carefully to an *even* yellow heat and placing it over a round hole in a swage-block or other object. The pene end of the hammer is used, and the pounding done over the hole in the swage-block. As the metal in the center is forced downward by the blow of the hammer, the swage-block prevents the material at the sides from following and is gradually worked into a bowl shape.

Fig. 119 shows the position of the block and the piece when forging.

The bowl being shaped properly, the lips should be formed, and the top of the bowl ground off true.

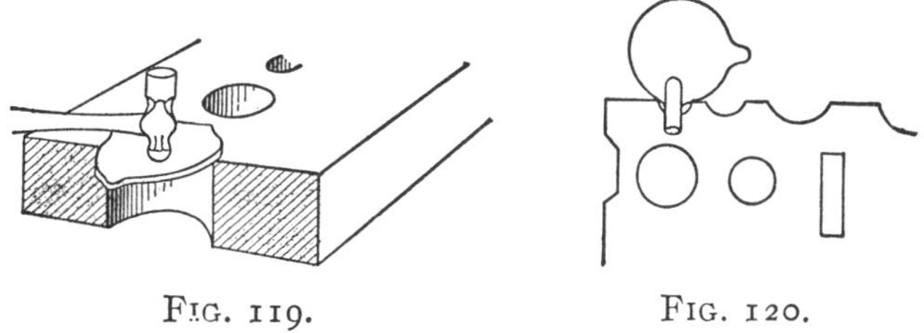

FIG. 119. FIG. 120.

The lips may be formed by holding the part where the lips are to be against one of the smaller grooves in the side of the swage-block, and driving it into the groove by placing a small piece of round iron on the inside of the bowl as shown in Fig. 120.

For a ladle with a bowl $3\frac{1}{2}''$ in diameter, the diameter of the circle, cut from the flat stock, should be about $4''$, as the edges of the piece draw in somewhat. Stock for other sizes should be in about the same proportion. Stock should be about $\frac{1}{8}''$ thick.

Machine-steel should be used for making the bowl. If ordinary wrought iron is used, the metal is liable to split.

Bowls.—Bowls, and objects of similar shape, may be made in the manner described above, but care must be used not to do too much hammering in the center of the stock, as that is the part most liable to be worked too thin.

Chain-stop.—The chain-stop, shown in Fig. 121, will serve as an example of a very numerous class of forgings; that is, forgings having a comparatively large projection on one side.

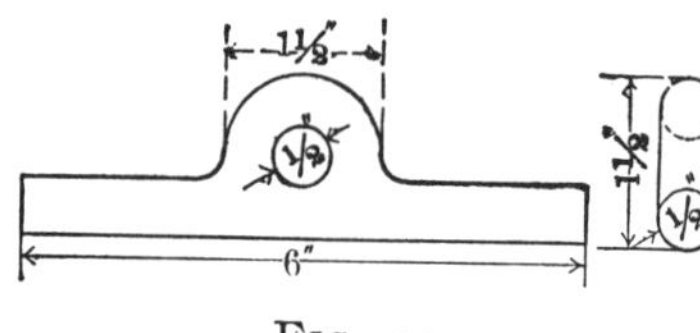

FIG. 121.

Care should be taken to select stock, for pieces of this sort, that will work into the proper shape with the least effort. The stock should be as thick as the thickest part of the forging, and as wide as the widest part. Stock, in this particular case, should be ½″ × 1½″.

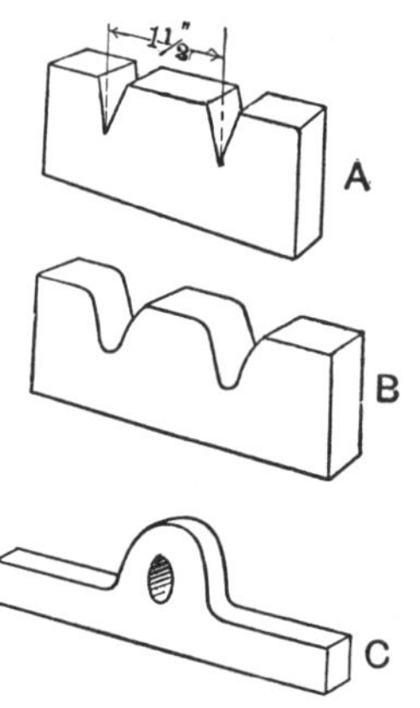

FIG. 122.

The different steps in making the forging are shown in Fig. 122. First two cuts are made 1½″ apart, as shown at *A*; then these cuts are widened out with a fuller, *B*. The ends are then forged out square, as at *C*. To finish the piece the hole is punched and rounded and the ends finished round.

When the fuller is used it should be held slightly slanting, as shown in Fig. 123.

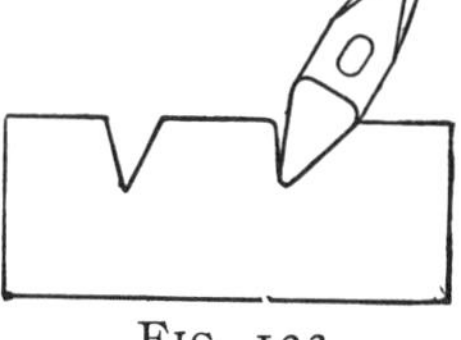
FIG. 123.

This forces the metal toward the central part and leaves a more nearly square shoulder, in place of the slanting shoulder that would be left were the fuller to be held exactly upright.

CHAPTER VI.

CALCULATION OF STOCK; AND MAKING OF GENERAL FORGINGS.

Stock Calculations for Forged Work.—When calculating the amount of stock required to make a forging, when the stock has its original shape altered, there is one simple rule to follow: Calculate the volume of the forging, add an allowance for stock lost in forging, and cut a length of stock having the total volume. In other words, the forging contains the same amount, or volume, of metal, no matter in what shape it may be, as the original stock; an allowance of course being made

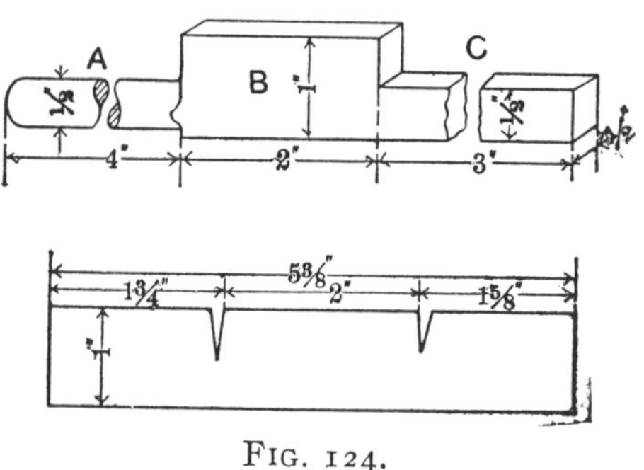

Fig. 124.

for the slight loss by scaling, and for the parts cut off in making.

Take as an example the forging shown in Fig. 124, to determine the amount of stock required to make the piece. This forging could be made in much the same way as the chain-stop. A piece of straight stock would be used and two cuts made and widened with a fuller, in the manner shown in Fig. 125. The ends on either side of the cuts

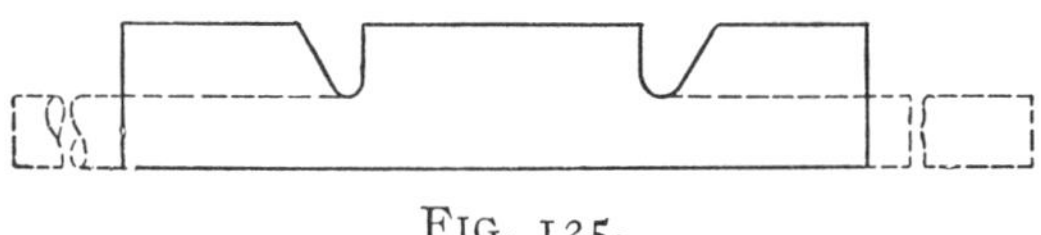

FIG. 125.

are then drawn down to size, as shown by the dotted lines, the center being left the size of the original bar. The stock should be $\frac{1}{2}'' \times 1''$, as these are the dimensions of the largest parts of the forging. For convenience in calculating the forging may be divided into three parts: the round end *A*, the central rectangular block *B*, and the square end *C*.

The block *B* will of course require just 2″ of stock.

The end *C* has a volume of $\frac{1}{2}'' \times \frac{1}{2}'' \times 3'' = \frac{3}{4}$ of a cubic inch.

The stock $(\frac{1}{2}'' \times 1'')$ has a volume of $\frac{1}{2}'' \times 1'' \times 1'' = \frac{1}{2}$ of a cubic inch for each inch of length.

To find the number of inches of stock required to make the end *C*, the volume of this end ($\frac{3}{4}$ cubic inch) should be divided by the volume of one inch of stock (or $\frac{1}{2}$ cubic inch). Thus, $\frac{3}{4} \div \frac{1}{2} = 1\frac{1}{2}''$.

It will therefore require $1\frac{1}{2}''$ of stock to make the end *C*; with allowance for scaling, $1\frac{5}{8}''$.

The end *A* is really a round shaft, or cylinder, 4″ long and $\frac{1}{2}''$ in diameter. To find the volume of a cylinder, multiply the square of half the diameter by $3^1/_7$, and then multiply this result by the length of the cylinder.

The volume of *A* would be $^1/_4 \times {}^1/_4 \times 3^1/_7 \times 4 = {}^{11}/_{14}$. And the amount of stock required to make *A* would be $^{11}/_{14} \div {}^1/_2 = 1^4/_7''$ in length, which is practically equal to $1^5/_8$. To the above amount of stock must be added a small amount for scaling, allowing altogether about $1^3/_4''$.

The stock needed for the different parts of the forging is as follows:

Round shaft *A*	$1\frac{3}{4}''$
Block *B*	$2''$
Square shaft *C*	$1\frac{5}{8}''$
Total.........................	$5\frac{3}{8}''$

First taking a piece of stock $\frac{1}{2}'' \times 1'' \times 5\frac{3}{8}''$, the cuts would be made for drawing out the ends as shown in Fig. 125.

In such a case as the above it is not always necessary to know the *exact* amount of stock to cut. What is known to be more than enough stock to make the forging could be taken, the central block made the proper dimensions, the extra metal worked down into the ends, and then trimmed off to the proper length. There are frequently times, however, when the amount of material required must be calculated accurately.

Take a case like the forging shown in Fig. 126. Here is what amounts to two blocks, each $2'' \times 4'' \times 6''$, connected by a round shaft, $2''$ in diameter.

To make this, stock 2″ thick and 4″ wide should be used, starting by making cuts as shown in Fig.

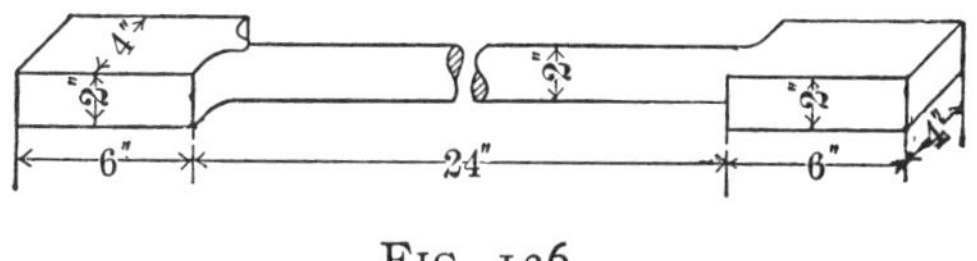

FIG. 126.

127, and drawing down the center to 2″ round. It is of course necessary to know how far apart to

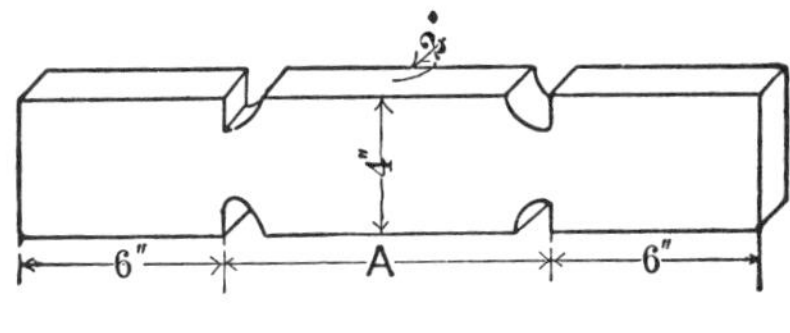

FIG. 127.

make the cuts when starting to draw down the center.

The volume of a cylinder 2″ in diameter and 24″ long would be $1'' \times 1'' \times 3^1/_7{}'' \times 24'' = 75^3/_7$ cubic inches, which may be taken as $75^1/_2$ cubic inches. For each inch in length the stock would have a volume of $4'' \times 2'' \times 1'' = 8$ cubic inches. Therefore it would require $75^1/_2 \div 8 = 9^7/_{16}$ inches of stock to form the central piece; consequently the distance between cuts, shown at *A* in Fig. 127, would have to be $9^7/_{16}{}''$. Each end would require 6″ of stock, so the total stock necessary would be $6'' + 6'' + 9^7/_{16}{}'' = 21^7/_{16}{}''$.

Any forging can generally be separated into several simple parts of uniform shape, as was done above, and in this form the calculation can be easily made, if it is always remembered that the *amount of metal* remains the same, and in forging, merely the shape, and not the volume, is altered.

Weight of Forgings.—To find the weight of any forging, the volume may first be found in cubic inches, and this volume multiplied by .2779, the weight of wrought iron per cubic inch. (If the forging is made of steel, multiply by .2836 in place of .2779.) This will give the weight in pounds.

Below is given the weight of both wrought and cast iron and steel, both in pounds per cubic inch and per cubic foot.

	Lbs. per Cu. Ft.	Lbs. per Cu. In.
Cast iron weighs.......	450	.2604
Wrought iron weighs..	480	.2779
Steel weighs...........	490	.2936

Suppose it is required to find the weight of the forging shown in Fig. 124. We had a volume in *A* of $^{11}/_{14}$ cubic inch, in *C* of $^{3}/_{4}$ cubic inch, and in *B* of 1 cubic inch, making a total of $2\,^{15}/_{28}$ cubic inches. If the forging were made of wrought iron, it would weight $2\,^{15}/_{28} \times .2779 = .7$ of a pound.

The forging shown in Fig. 126 has a volume in each end of 48 cubic inches, and in the center of $75\frac{3}{7}$ cubic inches, making a total of $171\frac{3}{7}$ cubic inches, and would weigh, if made of wrought iron, 47.64 pounds.

A much quicker way to calculate weights is to use a table such as is given on page 250. As steel is now commonly used for making forgings, this table is figured for steel. The weight given in the table is for a bar of steel of the dimensions named and one foot long. Thus a bar 1″ square weighs 3.402 lbs. per foot, a bar $3\frac{1}{2}''\times 1''$ weighs 11.9 lbs. per foot, etc.

To calculate the weight of the forging shown in Fig. 126, proceed as follows: Each end is $2''\times 4''$ and 6″ long, so, as far as weight is concerned, equal to a bar $4''\times 2''$ and 12″ long. From the table, a bar $4''\times 1''$ weighs 13.6 lbs. for each foot in length; so a bar $4''\times 2''$, being twice as thick, would weigh twice as much, or 27.2 lbs., and as the combined length of the two ends of the forging is one foot, this would be their weight. The table shows that a bar 2″ in diameter weighs 10.69 lbs. for every foot in length; consequently the central part of the forging, being 2 ft. long, would weigh 10.69×2, or 21.38 lbs. The total weight of the entire forging would be 48.58 lbs. (This seems to show a difference between this weight and the weight as calculated before, but it must be remembered that before the weight was calculated for wrought iron, while this calculation was made for steel.)

Finish.—Some forgings are machined, or "finished," after leaving the forge-shop. As the drawings are always made to represent the finished work, and give the finished dimensions, it is necessary to make an allowance for this finishing when making the forging, and all parts which have to be "finished," or "machined," must be left with extra metal to be removed in finishing.

The parts required to be finished are generally marked on the drawing; sometimes the finished surfaces have the word FINISH marked on them; sometimes the finishing is shown simply by the symbol *f*, as used in Fig. 128, showing that the shafts and pin only of the crank are to be finished.

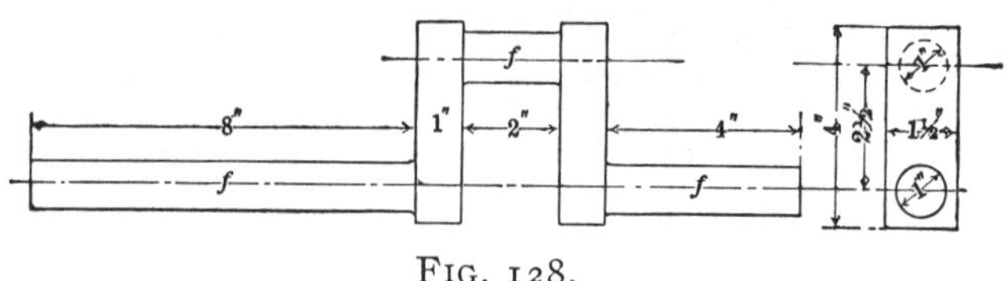

FIG. 128.

When all surfaces of a piece are to be finished the words FINISH-ALL-OVER are sometimes marked on the drawing.

The allowance for finish on small forgings is generally about $^1/_{16}''$ on each surface; thus if a block were wanted to finish $4'' \times 2'' \times 1''$, and $^1/_{16}''$ were allowed for finishing, the dimensions of the forging should be $4\frac{1}{8}'' \times 2\frac{1}{8}'' \times 1\frac{1}{8}''$.

On a forging like Fig. 126, about $\frac{1}{8}''$ allowance should be made for finish, if it were called for; thus the diameter of the central shaft would be $2\frac{1}{4}''$, the thickness of the ends $2\frac{1}{4}''$, etc. On larger work $\frac{1}{4}''$ is sometimes allowed for machining.

The amount of finish allowed depends to a large extent on the way the forging is to be finished. If it is necessary to finish by filing the forging should be made as nearly to size as possible, and having a very slight amount for finish, $^1/_{32}''$, or even $^1/_{64}''$, being enough in some cases.

It is of course necessary to take this into account when calculating stock, and the calculation made for the forging with the allowance for finish added to the drawing dimensions and not simply for the finished piece.

Crank-shafts. — There are several methods of forging crank-shafts, but only the common commercial method will be given here.

When forgings were mostly made of wrought iron, cranks were welded up of several pieces. One piece was used for each of the end shafts, one piece for each cheek, or side, and another piece for the crank-pin. Mild-steel cranks are now more universally used and forged from one solid piece of stock. The drawing for such a crank is given in Fig. 128; finish to be allowed only as shown, that is, only on crank-pin and shafts. The forgings, as made, will appear like the outlines in Fig. 129. The metal in the throat of the crank is generally removed by drilling a line of holes and then sawing slots where the sides of the crank cheeks should come, as shown by the dotted lines in Fig. 129.

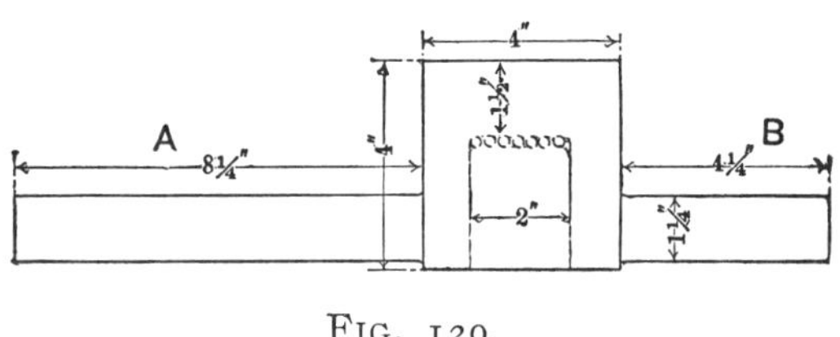

FIG. 129.

The central block is then easily knocked out. This drilling and sawing are done in the machine-shop. This throat can be formed by chopping out the surplus metal with a hot chisel, but on small cranks, such as here shown, it is generally cheaper in a well-equipped shop to use the first method.

The first step is to calculate the amount of stock required. Stock $1\frac{1}{2}'' \times 4''$ should be used. The ends, *A* and *B*, should be left $1\frac{1}{4}''$ in diameter to allow for finishing. The end *A* contains 10.13 cubic inches. Each inch of stock contains 6 cubic inches. It would therefore require 1.7″ of stock to form this end provided there were no waste from scale in heating. This waste does take place, and must be allowed for, so it will be safe to take about 2″ of stock for this end. *B* contains 5.22 cubic inches, and would require .87″ of stock without allowance for scale. About $1\frac{1}{8}''$ should be taken. The stock should then be $7\frac{1}{4}''$ long. The first step is to make cuts $1\frac{1}{8}''$ from one end and 2″ from the other, and widen out these cuts with a fuller, as shown in Fig. 130.

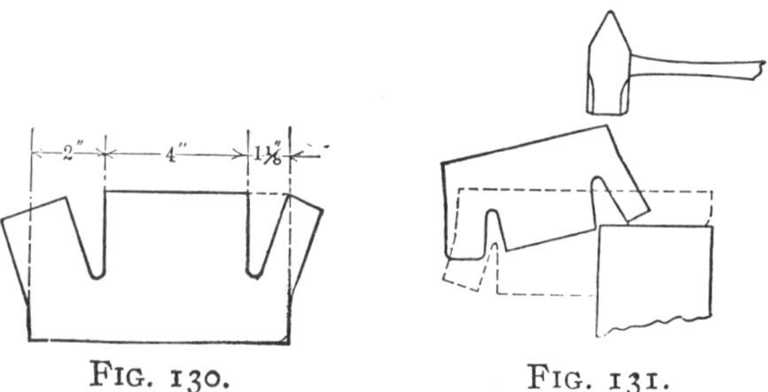

FIG. 130. FIG. 131.

These ends are then forged out round in the manner illustrated in Fig. 131. The forging should be placed over the corner of the anvil in the position shown, the blows striking upon the corner of the piece as indicated. As the end gradually straightens out, the other end of the piece is slowly raised into the position shown by the dotted lines and the shaft hammered down round and finished up between swages.

Care must be taken to spread the cuts properly before drawing down the ends, otherwise a bad cold-shut will be formed. If the cuts are left without spreading, the metal will act somewhat after the manner shown in Fig. 132. The top part of the bar, as it is worked down, will gradually fold over, leaving, when hammered down to size, a bad cold-shut, or crack, such as illustrated in Fig. 132. When the metal starts to act this way, as shown by the upper sketch in 132, the fault may be remedied by trimming off the corner along the dotted line. This must always be done as soon as any tendency to double over is detected.

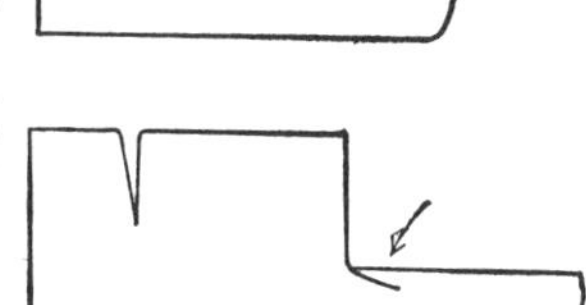

FIG. 132.

Double-throw Cranks.—Multiple-throw cranks are

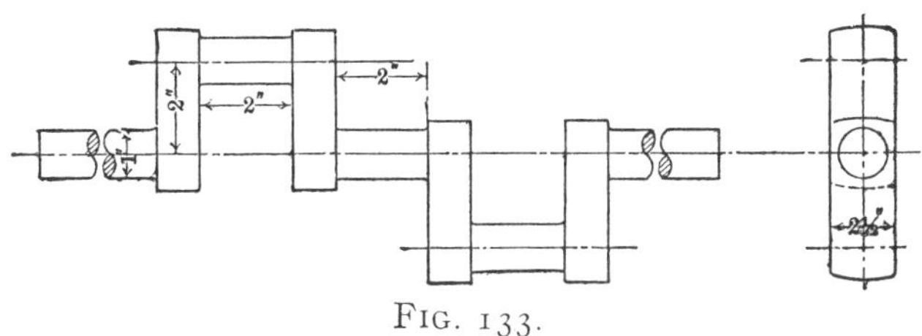

FIG. 133.

first forged flat, rough turned, then heated and twisted into shape.

The double-throw crank, shown in Fig. 133,

would be first forged as shown in Fig. 134; the parts shown dotted would then be cut out with the drill and saw, as described above.

After the pins and shafts have been rough turned —that is, turned round, but left as large as possi-

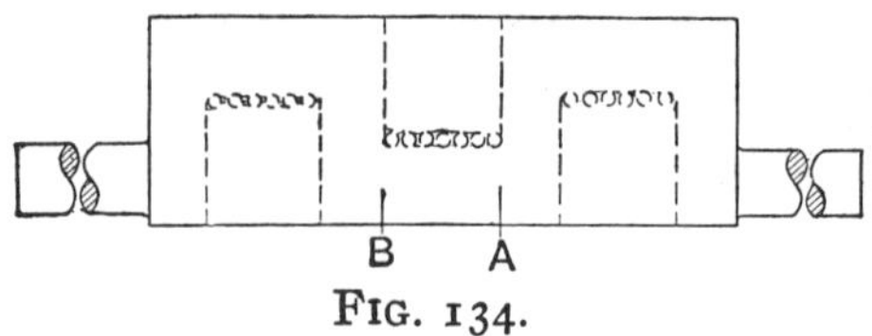

FIG. 134.

ble—the crank is returned to the forge-shop, where it is heated red-hot and twisted into the finished shape.

When twisting, the crank is gripped just to one side of the central bearing, as shown by the dotted line *A*. This may be done with a vise or wrench, if the crank is small, or the crank may be placed on the anvil of a steam-hammer and the hammer lowered down on it to hold it in place.

The other end of the crank is gripped on the line *B* and twisted into the required shape.

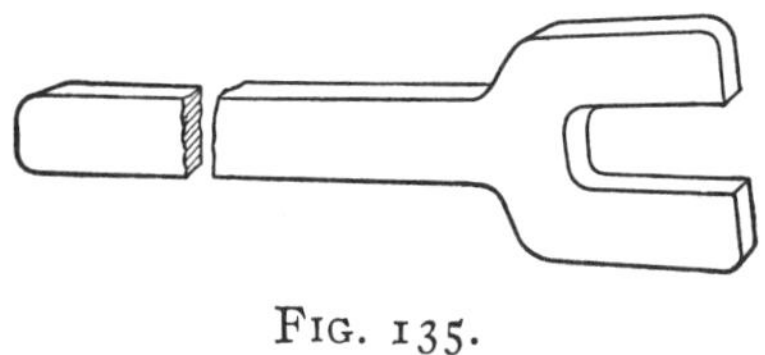

FIG. 135.

A wrench of the shape shown in Fig. 135 is very convenient for doing work of this character. It may be formed by bending a U out of flat stock, bent edgewise, and welding on a handle.

Three-throw Crank.—Fig. 136 shows what is known as a "three-throw" crank. The forging for

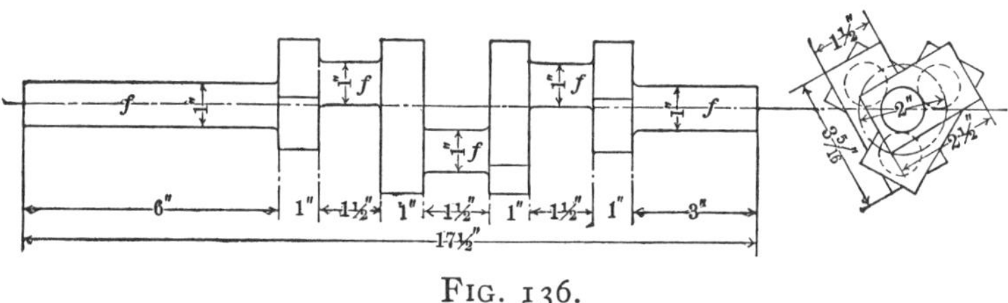

FIG. 136.

this is first made as shown by the solid lines in Fig. 137. The forging is drilled and sawed in the

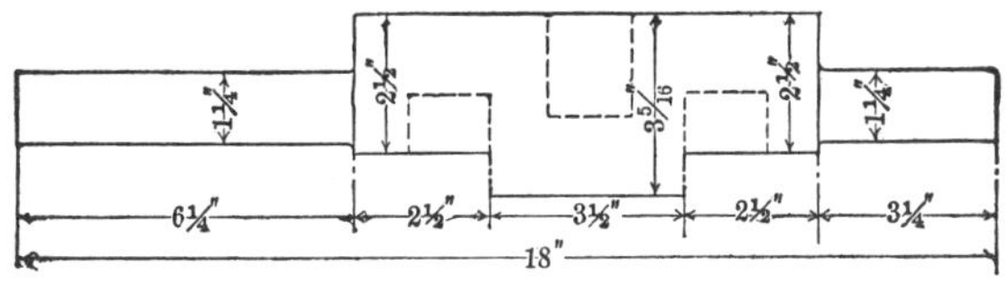

FIG. 137.

machine-shop to the dotted lines, and pins rough turned, being left as large as possible. The forging is returned to the forge-shop, heated, and bent into the shape of the finished crank. It is then sent to the machine-shop and finished to size. Four-throw cranks are also made in this manner.

The slots are sometimes cut out in the forge-shop with a hot chisel, but, particularly on small work, it is generally more economical to have them sawed out in the machine-shop. This is especially so of multiple-throw cranks, which must be twisted.

Knuckles.—There is a large variety of forgings which can be classed under one head—such forgings as the forked end of a marine connecting-rod, the knuckle-joints sometimes used in valve-rods, and others of this character, such as illustrated in Figs. 139, 140, 141, *E*.

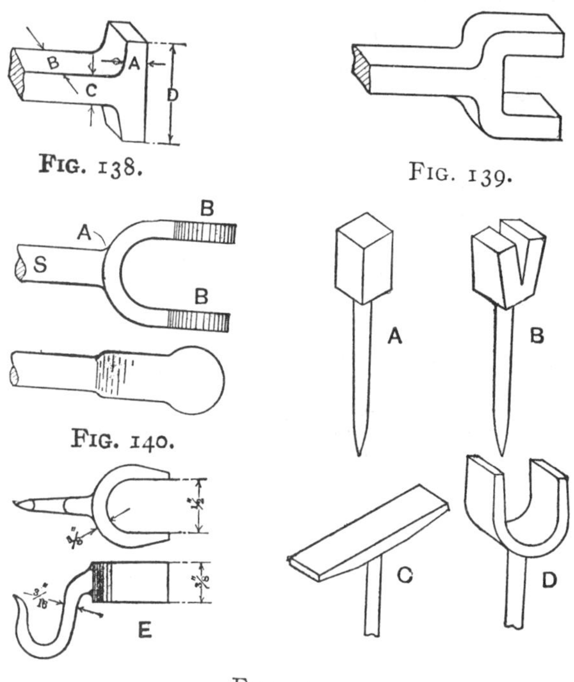

FIG. 138.

FIG. 139.

FIG. 140.

FIG. 141.

Connecting-rod End.—Fig. 138 shows the shaped end often used on the crank end of connecting-rods. The method of forming this is the same as the first step in forging the other pieces above mentioned.

The stock used for making this should be as wide

as *B* and somewhat more than twice as thick as *A*. The first step is to make two fuller cuts as shown at *A*, Fig. 142, using a top and bottom fuller and working in both sides at the same time. When working in both sides of a bar this way, it should be turned frequently, bringing first one side, then the other, uppermost. In this way the cuts will be worked to the same depth on both sides, while if the work is held in one position, one cut will generally be deeper than the other. After the cuts are made, the left-hand end of the bar is drawn out to the proper size and the right-hand end punched and split like *B*. Sometimes when the length *D*, Fig. 138, is comparatively short and the stock wide, instead of being punched and split, the end of the bar is cut out, as shown at *C*, Fig. 142, with a right angle or curved cutter.

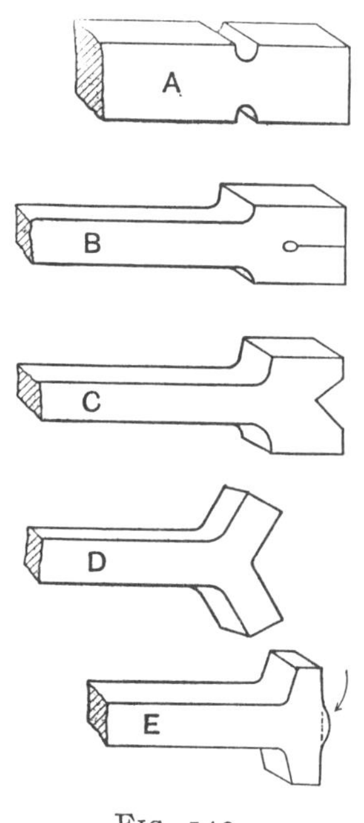

FIG. 142.

The split ends are spread out into the position shown at *D*, and drawn down to size over the corner of the anvil, in the manner illustrated in Fig. 143. These ends are then bent back into the proper position for the finished forging. Generally when the ends are worked out and bent back in this manner, a bump is left like that indicated by the arrow-point at *E*, Fig. 142. This should be trimmed off along the dotted line.

Knuckle.—The knuckle, Fig. 139, is started in exactly the same way, but after being forged out

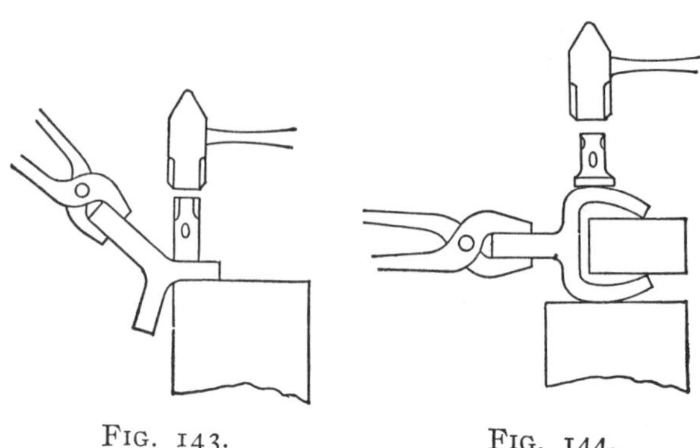

FIG. 143. FIG. 144.

straight, as above, the tips of these ends are bent down, forming a U-shaped loop of approximately the shape of the finished knuckle. A bar of iron of the same dimension at the inside of the finished knuckle is then inserted between the sides of the loop and the sides closed down flat over it, Fig. 144.

Forked-end Connecting-rod.—Fig. 140 is made in the same manner. The shaft *S* should be drawn down into shape and rounded up before the other end is split. After the split ends have been bent back straight, the shoulder *A* should be finished up with a fuller in the manner shown in Fig. 145. The rounded ends *B*–*B* should be formed before the piece is bent

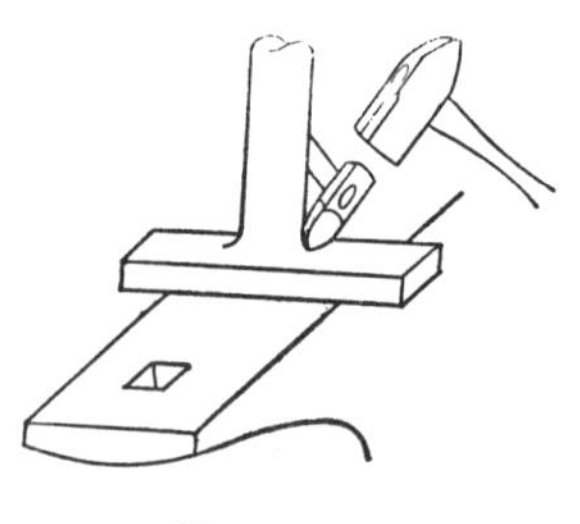

FIG. 145.

into shape. The final bending can be done over a cast-iron block of the right shape and size if the forging is a large one and several of the same kind are wanted.

Hook with Forked End.—Fig. 141, *E* is a forging which also comes in this general class. This is made from $\frac{3}{8}''$ square stock. The end of the bar is first drawn down to $^3/_{16}''$ round. This round end is put through the hole of a heading-tool, and the square part is split with a hot chisel, the cut widened out, and the sides hammered out straight on the tool. The different steps are shown in Fig. 141.

Wrench, Open-end.—Open-end wrenches of the general class shown in Fig. 146 may be made in

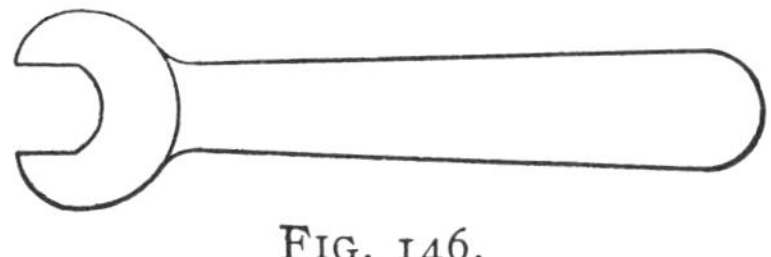

FIG. 146.

several different ways. It would be possible to make this by the same general method followed for making the forked end of the connecting-rod described above. Ordinary size wrenches are more easily made in the way illustrated in Fig. 147.

A piece of stock is used, wide enough and thick enough to form the head of the wrench. This is worked in on both sides with a fuller and the head rounded up as shown. A hole is then punched through the head and the piece cut out to form the opening, as shown by the dotted lines at *B*.

This wrench could also be made by bending up

a U from the proper size flat stock and welding on a handle.

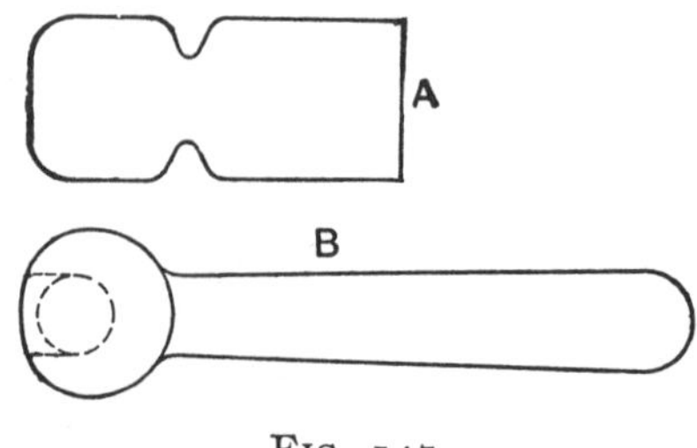

FIG. 147.

The solid-forged wrench is the more satisfactory.

Socket-wrench. — The socket-wrench, shown in Fig. 148, may be made in several ways. About the easiest, on "hurry-up" work, is the method shown in Fig. 149. Here a stub is shaped up the same size and shape as the finished hole is to be. A ring is bent up of thin flat iron and this ring welded around the stub.

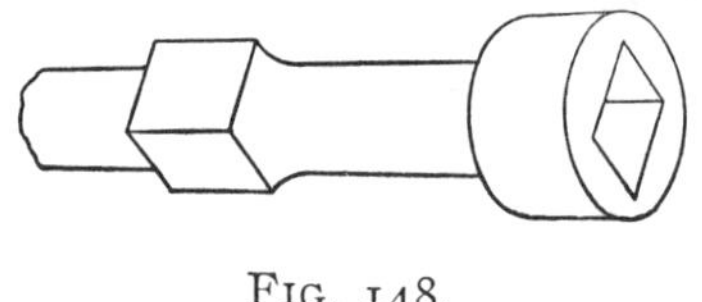

FIG. 148.

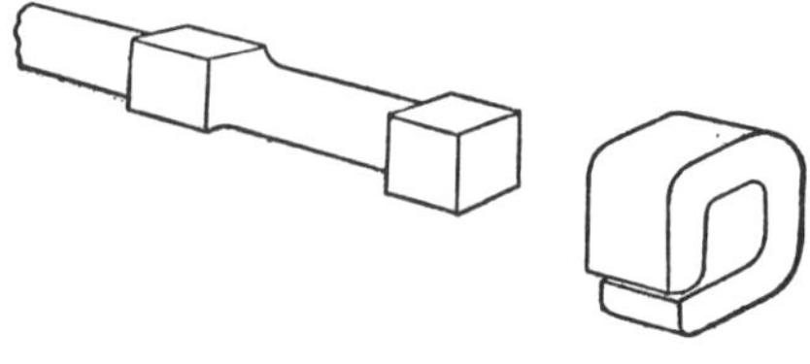

FIG. 149.

The width of the ring should of course be equal to the length of the hole plus the lap of the weld.

When finishing the socket, a nut or bolt-head the size the wrench is intended to fit should be placed in the hole and the socket finished over this between swages.

A better way of making wrenches of this sort is to make a forging having the same dimensions as the finished wrench, but with the socket end forged solid. The socket end should then be drilled to a depth slightly greater than the socket is wanted. The diameter of the drill should be, as shown in Fig. 150, equal to the shortest diameter of the hole.

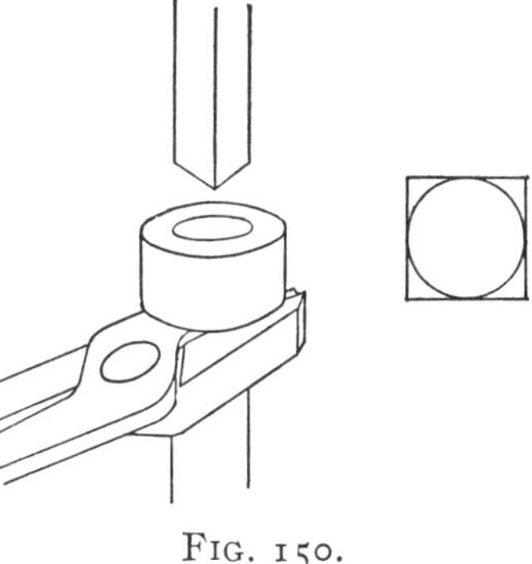

FIG. 150.

After drilling, the socket end is heated red-hot and a punch of the same shape as the intended hole driven into it. The end of the punch should be square, with the corners sharp. As the punch is driven in, the corners will shave off some of the metal around the hole and force it to the bottom of the hole, thus making it necessary to have the drilled hole slightly deeper than the socket hole is intended to finish.

While punching, the wrench may be held in a heading tool, or if the wrench be double-ended, in a pair of special tongs, as shown in Fig. 150.

Split Work.—There is a great variety of thin forgings, formed by splitting a bar and bending the split parts into shape. For convenience, these can be called split forgings.

Fig. 151 is a fair sample of this kind of work.

This piece could be made by taking two flat strips and welding them across each other, but, particu-

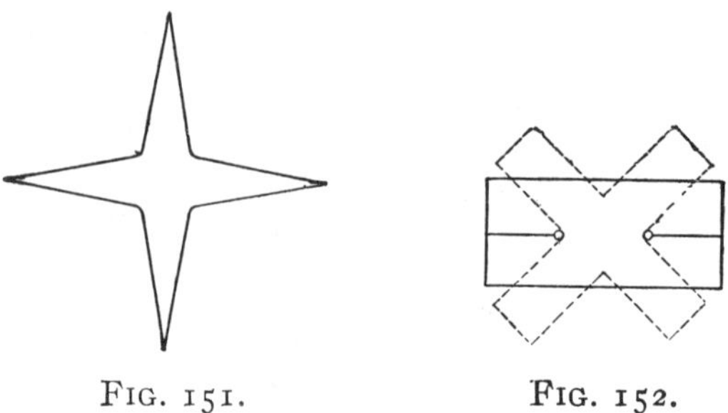

Fig. 151. Fig. 152.

larly if the work is very thin, this is rather a difficult weld to make.

An easier way is to take a flat piece of stock of the proper thickness and cut it with a hot chisel, as shown by the solid lines in Fig. 152. The four ends formed by the splits are then bent at right angles to each other as shown by the dotted lines, and hammered out pointed as required.

If machine steel stock is used, it is not generally necessary to take any particular precautions when

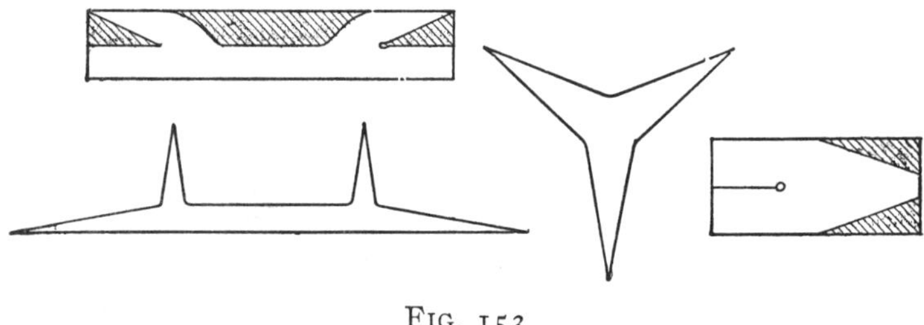

Fig. 153.

splitting the bar, but if the material used is wrought iron, it is necessary to punch a small hole through the bar where the end of the cut comes, to prevent the split from extending back too far.

Fig. 153 shows several examples of this kind of work. The illustrations show in each case the finished piece, and also the method of cutting the bar. The shaded portions of the bar are cut away completely.

Expanded or Weldless Eye. — Another forging of the same nature is the expanded eye in Fig. 154.

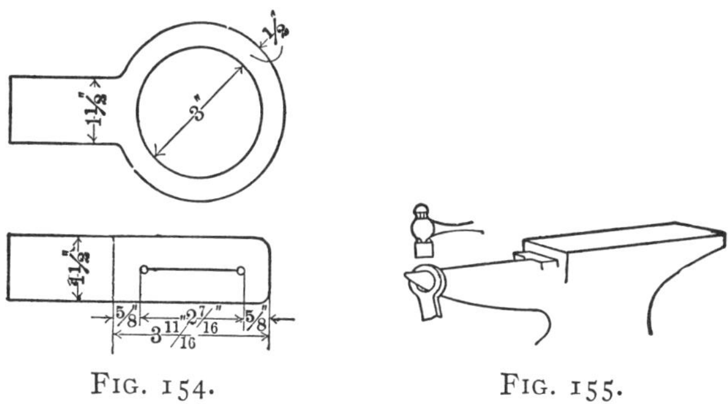

FIG. 154. FIG. 155.

To make this, a flat bar is forged rounding on the end, punched and split as shown. The split is widened out by driving a punch, or other tapering tool into it, and the forging finished by working over the horn of the anvil, as shown in Fig. 155.

If the dimensions of the eye are to be very accurate, it will be necessary to make a calculation for the length of the cut. This can be done as follows: Suppose the forging, for the sake of convenience in calculating, to be made up of a ring 3″ inside diameter and sides ½″ wide, placed on the end of a bar 1½″ wide. The first thing is to determine the area of this ring. To do this find the area of the outside circle and subtract from it the area of the inside circle. (Areas may be found in table, page 243.)

Area of outside circle	= 12.57	sq. in.
" " inside "	= 7.07	" "
" " ring	= 5.50	" "

The stock, being $1\frac{1}{2}''$ wide, has an area of $1\frac{1}{2}$ sq. in. for every inch in length, and it will take $3\frac{2}{3}''$ of this stock to form the ring, as we must take an amount of stock having the same area as the ring. This will be practically $3^{11}/_{16}''$.

The stock should be punched and split, as shown in Fig. 154. It will be noticed that the punch-holes are $\frac{5}{8}''$ from the end, while the stock is to be drawn to $\frac{1}{2}''$. The extra amount is given to allow for the hammering necessary to form the eye.

Weldless Rings.—Weldless rings can be made in the above way by splitting a piece of flat stock and expanding it into a ring, or they can be made as follows: The necessary volume of stock is first forged into a round flat disc and a hole is punched through the center. The hole should be large enough to admit the end of the horn of the anvil. The forging is then placed on the horn and worked to the desired size in the manner indicated in Fig. 155. Fig. 156 shows the different steps in the process—the disc, the punched disc, and the finished ring.

Rings of this sort are made very rapidly under the steam-hammer by a slight modification of this method. The discs are shaped and punched and then forged to size over a "mandril." A U-

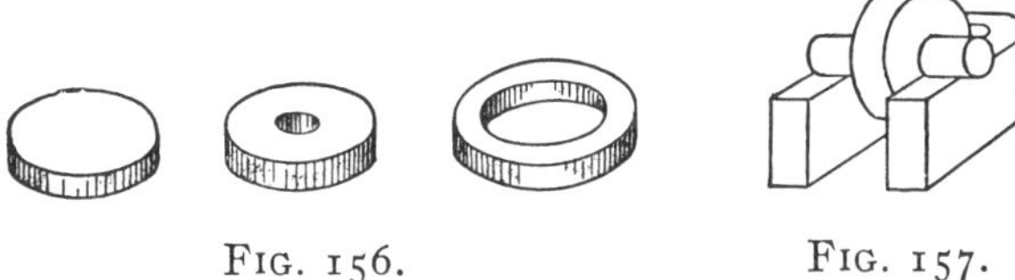

Fig. 156. Fig. 157.

shaped rest is placed on the anvil of the steam-hammer, the mandril is slipped through the hole in the disc and placed on the rest, as shown in Fig. 157. The blows come directly down upon the top side of the ring, it being turned between each two blows. The ring of course rests only upon the mandril. As the hole increases in size, larger and larger mandrils are used, keeping the mandril as nearly as possible the same size as the hole.

Forging a Hub, or Boss.—Fig. 158 is an example of a shape very often met with in machine forging: a lever, or some flat bar or shank, with a "boss"

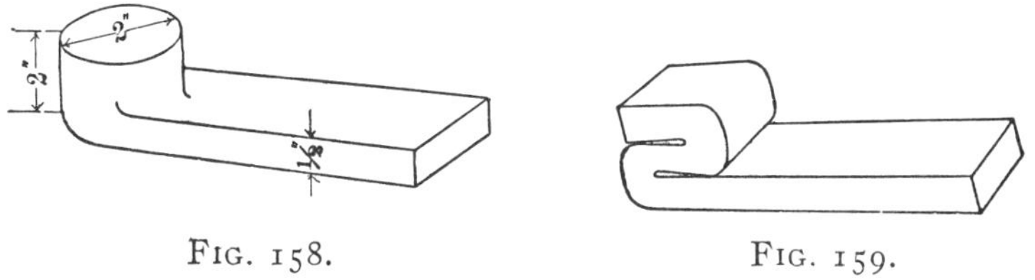

Fig. 158. Fig. 159.

formed on one end. This may be made in two ways—either by doubling over the end of the bar, as shown in Fig. 159, and making a fagot-weld of sufficient thickness to form the boss, or by taking a bar large enough to form the boss and drawing down the shank. The second method will be described, as no particular directions are necessary for the weld, and after welding up the end, the boss is rounded up in the same way in either case. The stock should be large enough to form the boss without any upsetting.

A bar of stock is taken, for the forging shown above, 2″ wide and 2″ thick. The first step is to make a cut about 2″ from the end, with a fuller, like *A*, Fig. 160.

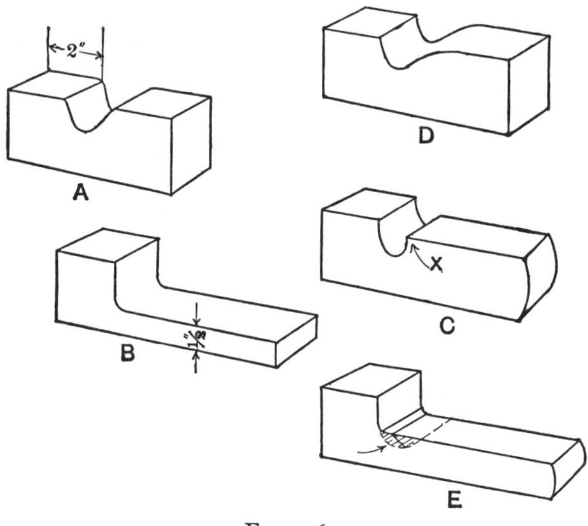

Fig. 160.

The stock, to the right of the cut, is then flattened down and drawn out to size, as shown at *B*. In drawing out the stock, certain precautions must be taken or a "cold-shut" will be formed close to the boss. If the metal is allowed to flatten down into shape like Fig. *C*, the corner at *X* will overlap, and work into the metal, making a crack in the work which will look like Fig. *E*. This is quite a common fault, and whenever a crack appears in a forging close to a shoulder, it is generally caused by something of this sort—that is, by some corner or part of the metal lapping over and cutting into the forging. When one of these cracks appears, the only way to remedy the evil is to cut it out as shown by the dotted lines in *E*. For this purpose a hot-chisel is sometimes used, with a blade formed like a gouge.

Fig. *D* shows the proper way to draw out the stock; the corner in question should be forged away from the boss in such a manner as to gradually widen the cut. The bar should now be rounded up by placing the work over the corner of the anvil, as shown in Fig. 161. First forge off the

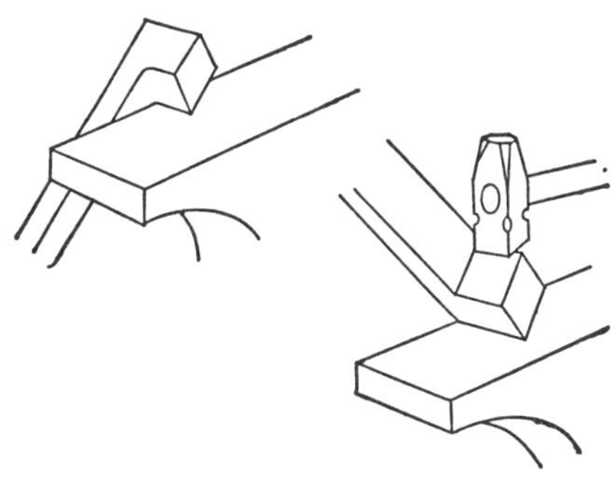

FIG. 161.

corners and then round up the boss in this way. To finish around the corner formed between the boss and the flat shank, a set-hammer should be used. Sometimes the shank is bent away from the boss to give room to work, and a set-hammer, or swage, used for rounding the boss as shown.

After the boss is finished, the shank is straightened. The boss should be smoothed up with a swage.

Ladle Shank.—The ladle shank, shown in Fig. 162, may be made in several ways. It is possible to make it solid without any welds, or the handle may be welded on a flat bar and the bar bent into a ring and welded, or the ring and handle may be forged in one piece and the ring closed together by welding. The last-mentioned method is as follows: The stock should be about 1″ square. It is necessary to make a rough calculation of the amount of this size stock required to form the ring of the shank. If the ring

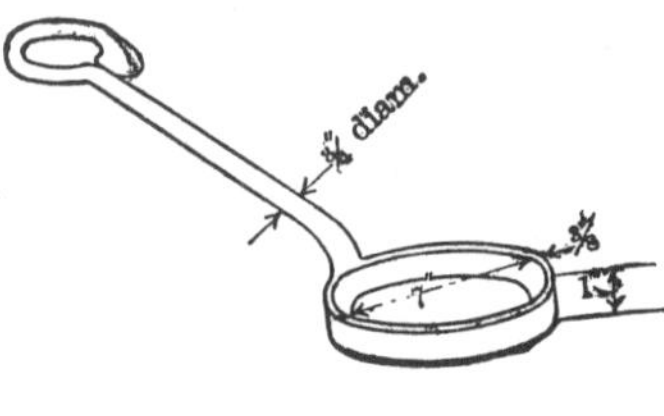

FIG. 162.

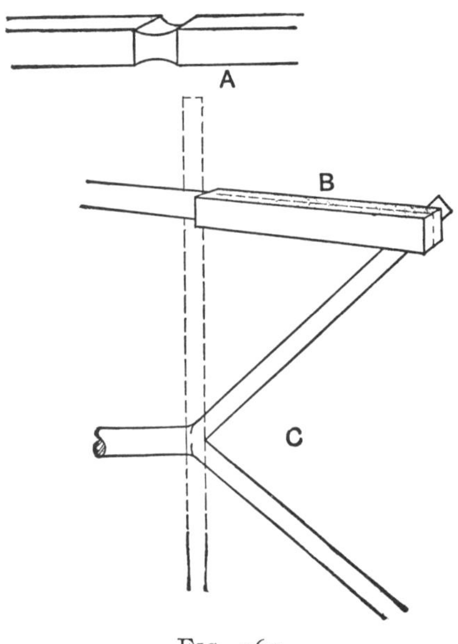

FIG. 163.

were made of $\frac{3}{8}'' \times 1''$ stock, about $23\frac{1}{4}''$ would be required; now as $1'' \times 1''$ stock is the same width and about two and one-half times as thick as $\frac{3}{8}'' \times 1''$ stock, every inch of the $1'' \times 1''$ will make about $2\frac{1}{2}''$ of $\frac{3}{8}'' \times 1''$, consequently about $9\frac{1}{2}''$ of the 1″ square will be required to form the ring.

A fuller cut is made around the bar, as shown at *A*, Fig. 163. This should be made about $9\frac{1}{2}''$ from the end of the bar. The left-hand end of the bar is drawn down to $\frac{3}{4}''$ in diameter to form the handle. If the work is being done under a steam or power hammer, enough stock may be drawn out to form the entire handle, but if working on the anvil, it will probably be more satisfactory to draw out only enough stock to make a "stub" 4″ or 5″ long. To this stub may be welded a round bar to form the handle.

After drawing out the handle, the $9\frac{1}{2}''$ square end of the stock is split, as shown by the dotted lines at *B*. These split ends are spread apart, as shown at *C*, forged into shape, and bent back to the position shown by the dotted lines.

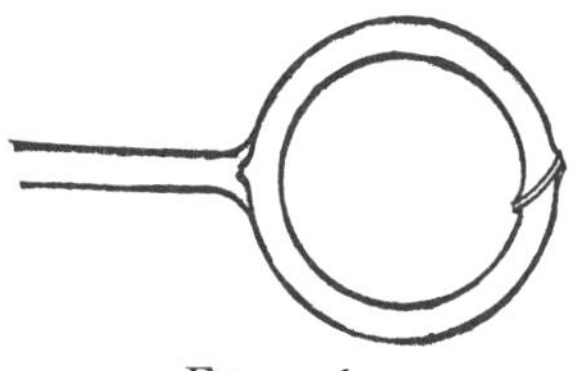

FIG. 164.

The ring is completed by cutting the ends to the proper length, scarfing, bending into shape, and welding, as indicated in Fig. 164.

If for any reason it is necessary to make a forging of this kind without a weld in the ring, it may be done by the method shown in Fig. 165. The split in this case should not extend to the end of the bar. About $\frac{5}{8}''$ or $\frac{3}{4}''$ of stock should be left uncut at the end. This split is widened out and the

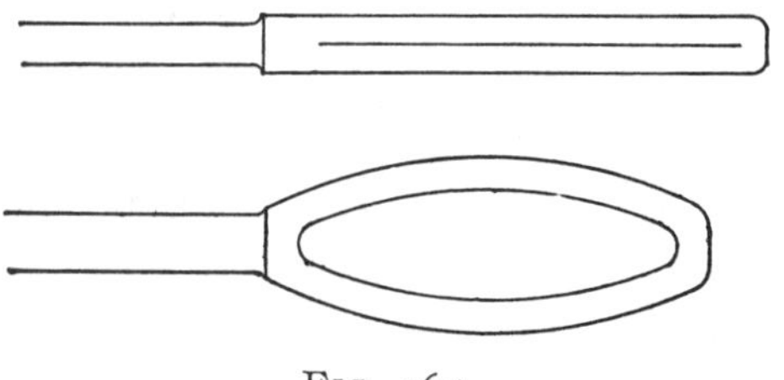

FIG. 165.

sides drawn down and shaped into a ring as desired.

Starting-lever.—The lever shown in Fig. 166 is a

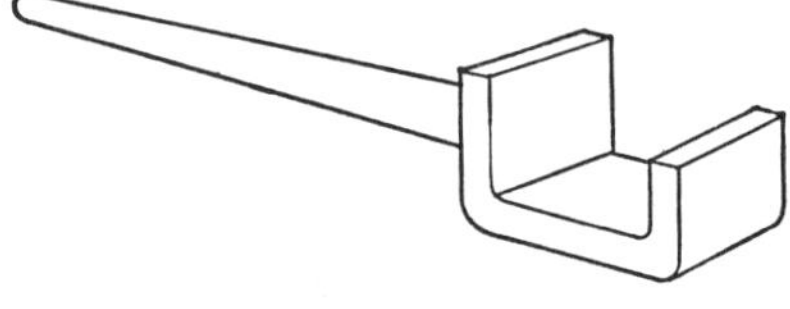

FIG. 166

shape sometimes used for levers used to turn the fly-wheels of engines or other heavy wheels by gripping the rim.

The method used in making the lever is shown in Fig. 167. The end is first drawn down round and the handle formed. The other end is then split, forged down to size, and bent at right angles to the handle. After trimming to the proper length, the flat ends are bent into shape.

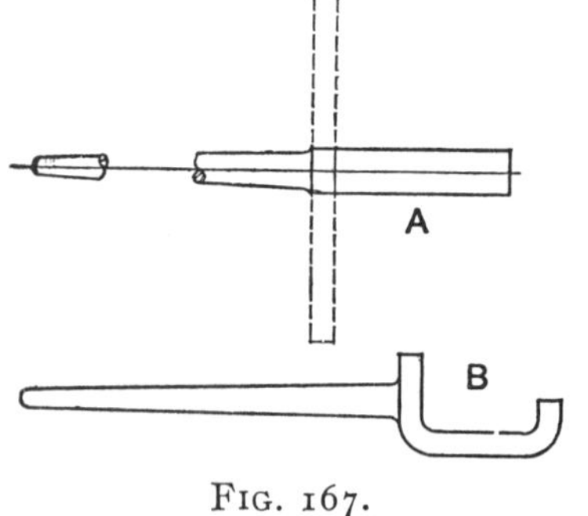

FIG. 167.

If this shaped end is very heavy, it may be necessary to forge it in the shape of a solid block, as shown in Fig. 168, and then either work in the depression shown by the dotted lines, with a fuller and set-hammers, or the dotted line may be cut out with a hot-chisel.

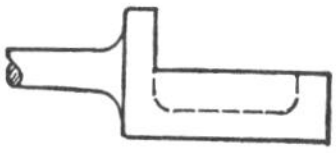

FIG. 168.

Moulder's Trowel.—The moulder's trowel shown in Fig. 169 gives an example of the method used in

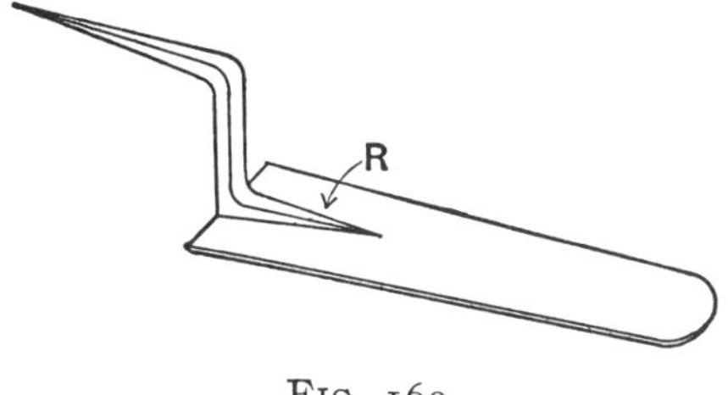

FIG. 169.

making forgings of a large class, forgings having a wide thin face with a stem, comparatively small, forged at one end.

The stock to be used for the trowel shown should be about $\frac{1}{4}'' \times 1''$. This is thick enough to allow for the formation of the ridge at *R*.

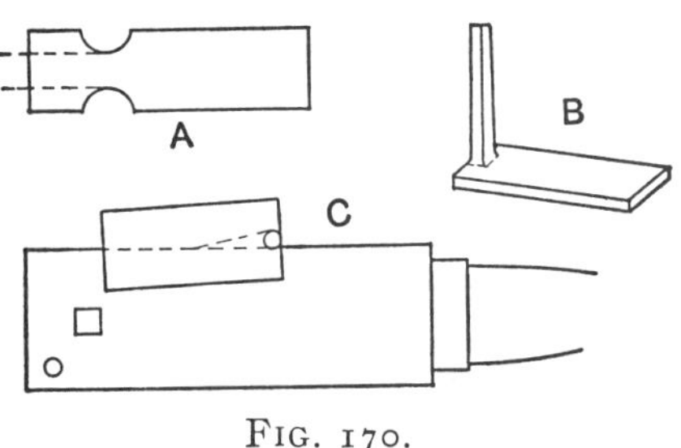

FIG. 170.

Fig. 170 shows the general method employed. The forging is started by making nicks like *A*, with the top and bottom fuller. One end is drawn down to form the tang for the handle. This should not be forged down pointed, as required when completed, but the entire length of handle should be forged square and about the size the largest part is required to finish to. The handle is then bent up at right angles, as at *B*, and the corner forged square in the same manner that the corner of a bracket is shaped up sharp and square on the outside.

After this corner is formed, the blade is drawn down to size on the face of the anvil.

When flattening out the blade, in order to leave the ridge shown at *R*, Fig. 169, the work should be held as shown at *C*, Fig. 170. Here the handle is held pointing down and against the side of the anvil. By striking directly down on the work, and covering the part directly over the edge of the anvil with the blows, all of the metal on the anvil will be flattened down, leaving the metal not resting on the anvil unworked. By swinging the piece around into a reverse position the other edge of the blade may be thinned down. If care be taken to hold the trowel in the proper position while thinning out the blade, a small triangular-shaped piece next the handle will be left thicker than the rest of the blade. This raised part will form the ridge shown at *R*, Fig. 169.

The same result may be obtained by placing the trowel, other side up, on the face of the anvil and using a set-hammer, or flatter, to thin out the blade.

Welded Brace.—Fig. 171 shows a form of brace, or bracket, sometimes used for holding swinging signs and for various other purposes.

The bracket in this case is made of round stock; but the same method may be followed in making one of flat or square material.

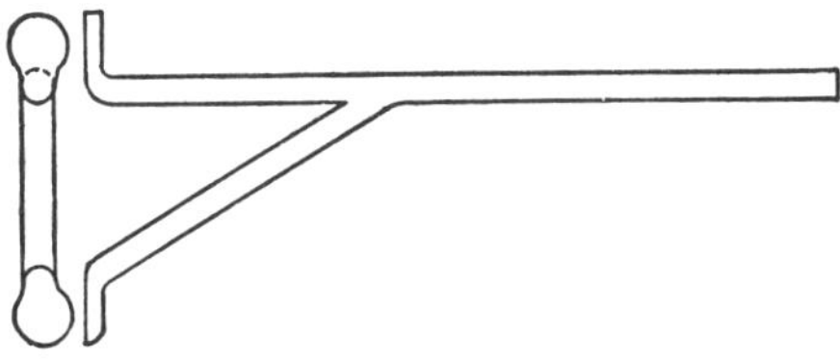

FIG. 171.

The stock is first scarfed on one end and this end doubled over, forming a loop, as shown in Fig. 172.

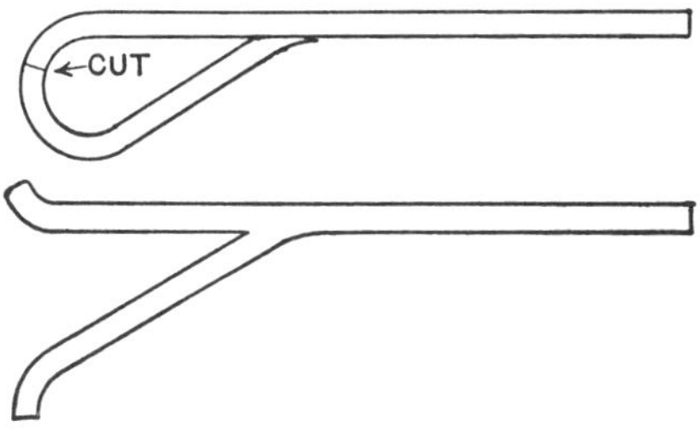

FIG. 172.

The loop is welded and then split, the ends straightened out and flattened into the desired shape as illustrated.

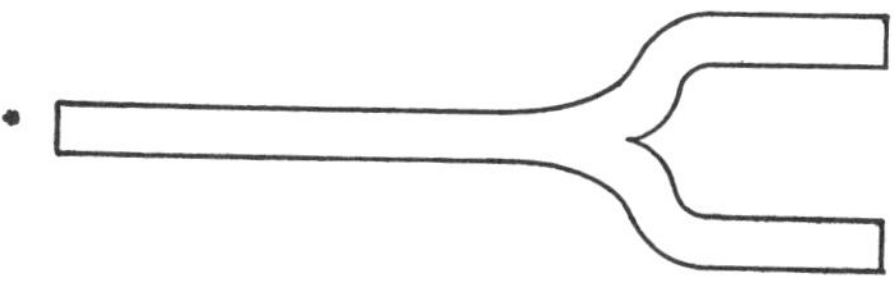

FIG. 173.

Welded Fork.—The welded fork, shown in Fig. 173, is made in the same way as the brace described above.

CHAPTER VII.

STEAM-HAMMER WORK.

General Description of Steam-hammer.—The general shape of small and medium steam-hammers is shown in Fig. 174. This type is known as a single-frame hammer.

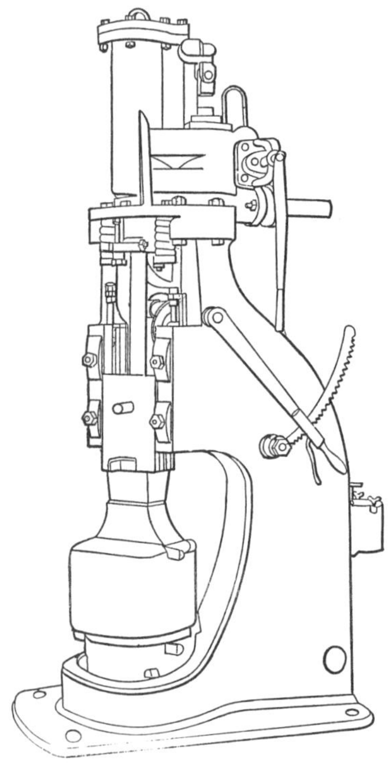

FIG. 174.

The size of a steam-hammer is determined by the weight of its falling parts; thus the term a 400-lb. hammer would mean that the total weight of the ram, hammer-die, and piston-rod was 400 lbs.

Steam-hammers are made in this general style from 200 lbs. up.

The anvil is entirely separate from the frame of the hammer, and each rests on a separate foundation.

The foundation for the frame generally takes the shape of two blocks of timber or masonry capped with timber—one in front and one behind the anvil block. The anvil foundation is placed between the two blocks of the frame foundation, and is larger and heavier.

The object of separating the anvil and frame is to allow the anvil to give under a heavy blow without disturbing the frame or its foundation.

Hammer-dies.—The dies most commonly used on steam-hammers have flat faces; the upper or hammer die being the same width, but sometimes shorter in length than the lower or anvil die.

Tool-steel makes the best dies, but chilled iron is also used to a very large extent. Sometimes, for forming work, even gray iron castings are used. Flat dies made of tool-steel are sometimes used without hardening. Dies made this way, when worn, may be faced off and used again without the bother of annealing and rehardening.

For special work the dies are made in various shapes, the faces being more or less in the shape of the work to be formed. When the die-faces are shaped to the exact form of the finished piece, the work is known as drop-forging.

Tongs for Steam-hammer Work.—The tongs used for holding work under the steam-hammer should be very carefully fitted and the jaws so shaped that they hold the stock on all sides. Ordinary flat-jawed tongs should not be used, as the work is liable to be jarred or slip out sideways.

Fig. 175 shows the jaws of a pair of tongs fitted

to square stock. Tongs for other shaped stock should have the jaws formed in a correspond ing way; that is, the inside of the jaws, viewed from the end, should have the same shape as the cross-section of the stock they are intended to hold, and should grip the stock firmly on *at least three sides.*

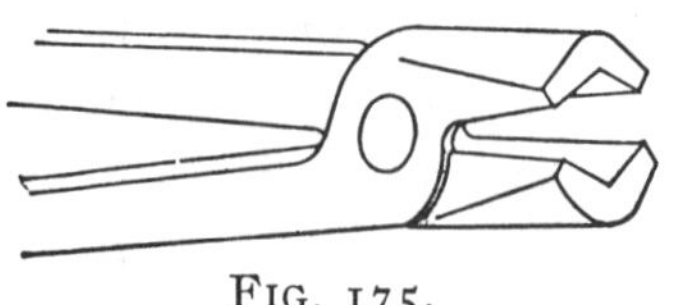

FIG. 175.

Flat-jawed tongs can be easily shaped as above in the manner shown in Fig. 176. The tongs are

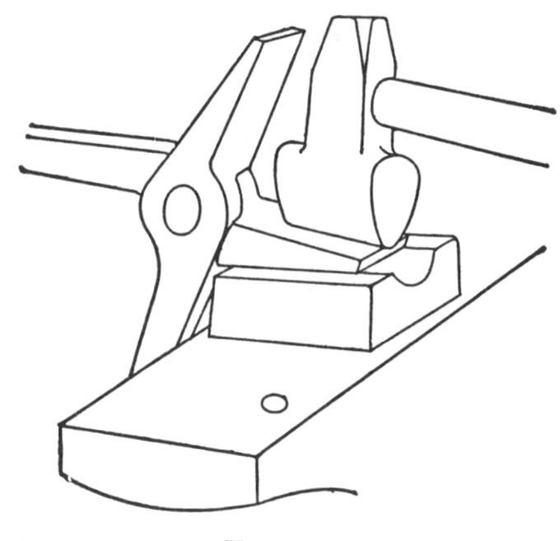

FIG. 176.

heated and held as shown, by placing one jaw, inside up, on a swage. The jaw is grooved or bent by driving down a top-fuller on it. After shaping the other jaw in the same way, the final fitting is done by inserting a short piece of stock of the proper size in the jaws and closing them down tightly over this by hammering.

When fitting tongs to round stock, the finishing

may be done between swages, the stock being kept between the jaws while working them into shape.

Tongs for heavy work should have the jaws shaped as shown in Fig. 177. When in use, tongs of this kind are held by slipping a link over the handles to force them together. On very large sizes, this link is driven on with a sledge.

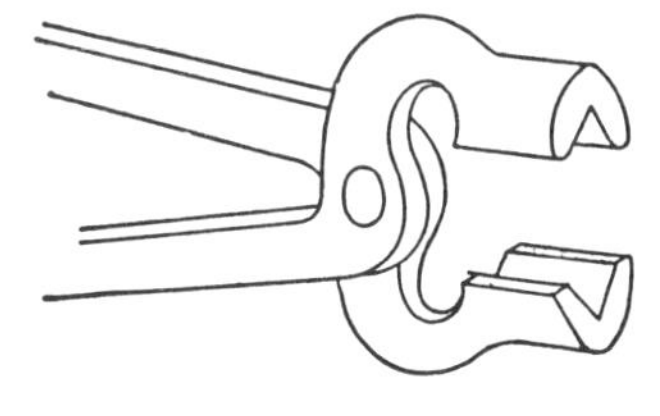

Fig. 177.

To turn the work easily, the link is sometimes made in the shape shown in Fig. 178, with a handle projecting from each end.

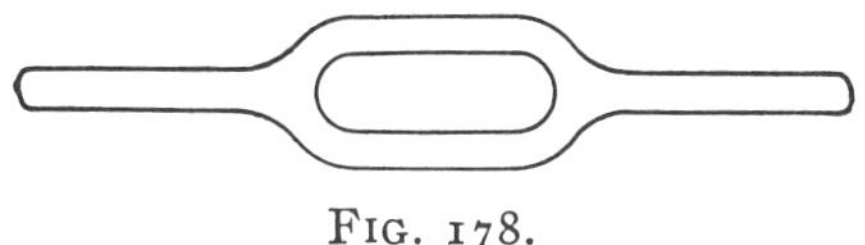

Fig. 178.

Hammer-chisels.—The hot-chisel used for cutting work under the hammer is shaped, ordinarily, like Fig. 179. This is sometimes made of solid tool-steel, and sometimes the blade is made of tool-steel and has a wrought-iron handle welded on. Fig. 180 shows the method of welding on the wrought-iron handle.

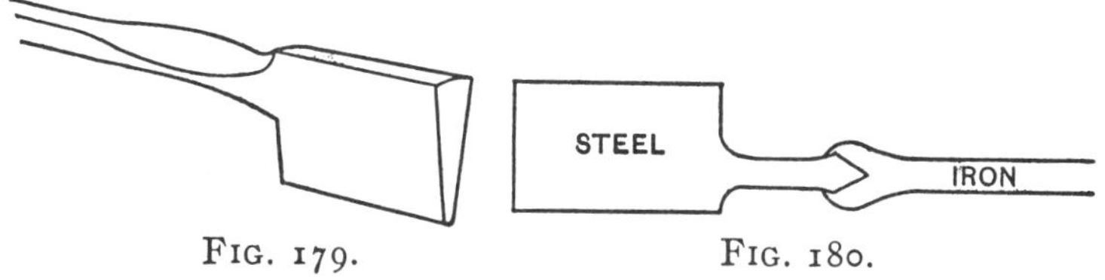

Fig. 179. Fig. 180.

The handle of the chisel, close up to the blade, is hammered out comparatively thin. This is to allow the blade to spring slightly without snapping off the handle. The hammer will always knock the blade into a certain position, and as the chisel is not always held in exactly the right way, this thin part of the handle permits a little "give" without doing any harm.

The force of the blow is so great when cutting, that the edge of the chisel must be left rather blunt. The edge should be square across, and not rounding. The proper shape is shown at *A*, Fig.

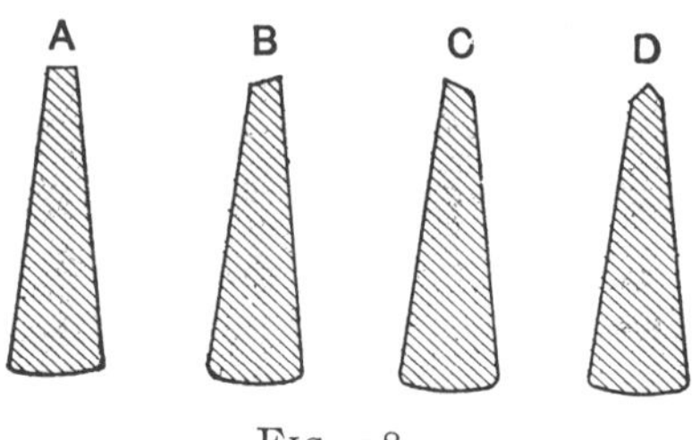

Fig. 181.

181. Sometimes for special work the edge may be slightly beveled, as at *B* or *C*, but should never be shaped like *D*.

Sometimes a bar is cut or nicked with a cold-chisel under the hammer. The chisel used is shaped like Fig. 182, being very flat and stumpy to resist the crushing effect of heavy blows. The three faces of the chisel are of almost equal width.

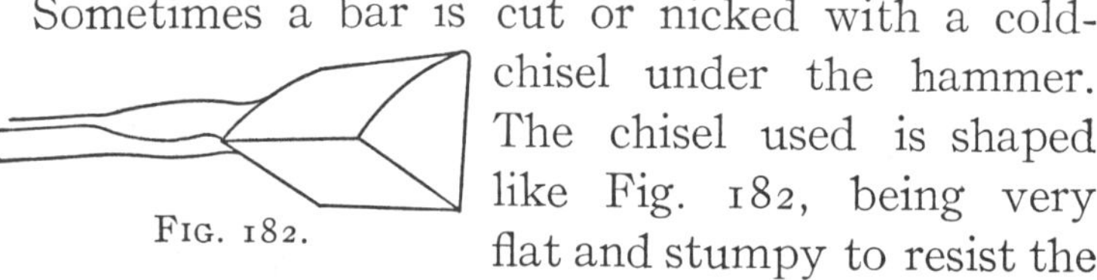

Fig. 182.

Cutting Hot Stock.—Hot cutting is done under the

steam-hammer in much the same way as done on the anvil.

If the chisel be held perfectly upright, as shown at *A*, Fig. 183, the cut end of the bar will be left

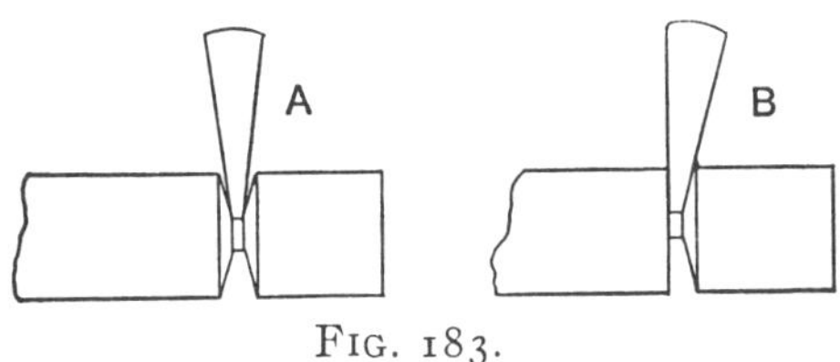

FIG. 183.

bulging out in the middle. When the end is wanted square the cut should be started with the chisel upright, but once started, the chisel should be very slightly tipped, as shown at *B*. When cutting work this way the cut should be made about half way through from all sides. When cutting off large pieces of square stock the chisel should be driven nearly through the bar, leaving only a thin strip of metal, $\frac{1}{8}''$ or $\frac{1}{4}''$ thick, joining the two pieces, *A*, Fig. 184. The bar is then turned over

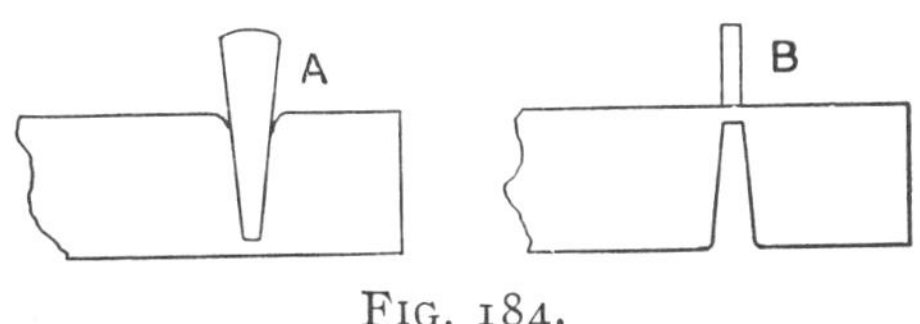

FIG. 184.

on the anvil and a thin bar of steel laid directly on top of this thin strip, as shown at *B*, Fig. 184. One hard blow of the hammer sends the thin bar of steel between the two pieces and completely cuts out the thin connecting strip of metal. This leaves the ends of both pieces smooth, while if the chisel is used for cutting on both sides, the end of one piece will be smooth and the other will have a fin left on it.

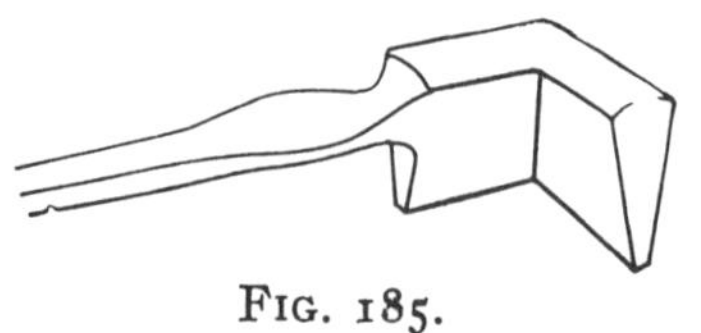

FIG. 185.

For cutting up into corners on the ends of slots bent cutters are sometimes used; such a cutter is shown in Fig. 185. These cutters are also made curved, and special shapes made for special work.

General Notes on Steam-hammer.—When working under the hammer, great care should always be taken to be sure that everything is in the proper position before striking a blow. The work *must* rest *flat* and *solid* on the anvil, and the part to be worked should be held as nearly as possible below the center of the hammer-die; if the work be done under one edge or corner of the hammer-die, the result is a "foul" blow, which has a tendency to tip the ram and strain the frame.

When tools are used, they should always be held in such a way that the part of the tool touching the work is directly below the point of the tool on which the hammer will strike. Thus, supposing a piece were being cut off under the hammer, the chisel should be held exactly upright, and directly under the center of the hammer, as shown at *A*, Fig. 186. In this way a fair cut is made. If the chisel were not held upright, but slantingly, as shown at *B*, the result of the blow would be as shown by the dotted lines, the chisel would be turned over and knocked flat, and, in some cases, might be even thrown very forcibly from under the hammer.

When a piece is to be worked out to any great extent, the blows should be *heavy*, and the end of

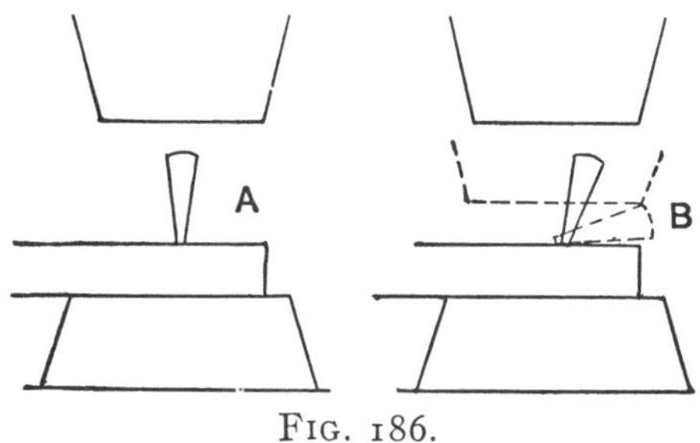

FIG. 186.

the stock being hammered should bulge out slightly, like *A*, Fig. 187, showing that the metal is being

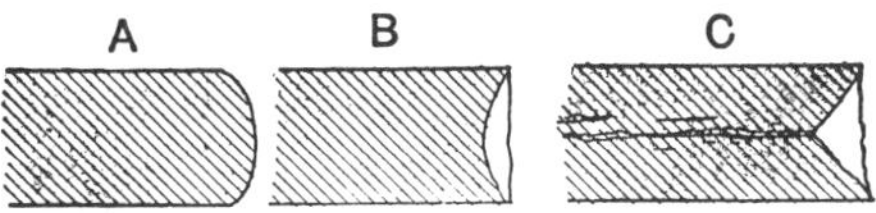

FIG. 187.

worked clear through. If light blows are used the end of the piece will forge out convex, like *B*, showing that the metal on the outside of the bar has been worked more than that on the inside. If this sort of work is continued, the bar will split and work hollow in the center, like *C*.

Round shafts formed between flat dies are very liable to be split in this way when not carefully handled.

The faces of the hammer- and anvil-dies are generally of the same width, but not always the same length. Thus, when the hammer is resting on the anvil, the front and back sides of the two dies are in line with each other, while either one or both ends of the anvil-die project beyond the ends of the hammer-die.

This is not always the case, however, as in many hammers the faces of the two dies are the same shape and size.

Having one die face longer than the other is an advantage sometimes when a shoulder is to be formed on one side of the work only.

When a shoulder is to be formed on both sides of a piece the work should be placed under the hammer in such a way that the top die will work in one shoulder, while the bottom die is forming the other; in other words, the work should be done from the side of the hammer, where the edges of the dies are even, as shown in Fig. 188. If the shoulder is required on one side only, as in forging tongs, the

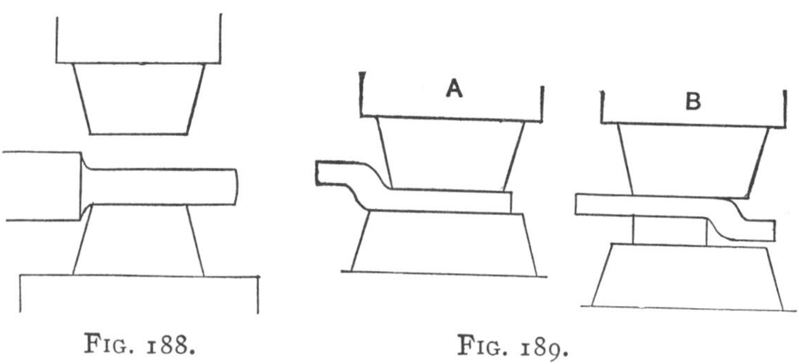

Fig. 188. Fig. 189.

work should be so placed as to work in the shoulder with the top die, while the bottom die keeps the under side of the work straight, as in Fig. 189, *A*.

The same object, a shoulder on one side only, may be accomplished by using a block, as shown at *B*, Fig. 189. The block may be used as shown, or the positions of work and block may be reversed and the work laid with flat side on the anvil and block placed on top.

This method of forming shoulders will be taken up more in detail in treating individual forgings.

Tools: Swages.—In general, the tools used in steam-hammer work, except in special cases, are very simple.

Swages for finishing work up to about 3″ or 4″ in diameter are commonly made as shown in Fig. 190. The two parts of the swage are held apart

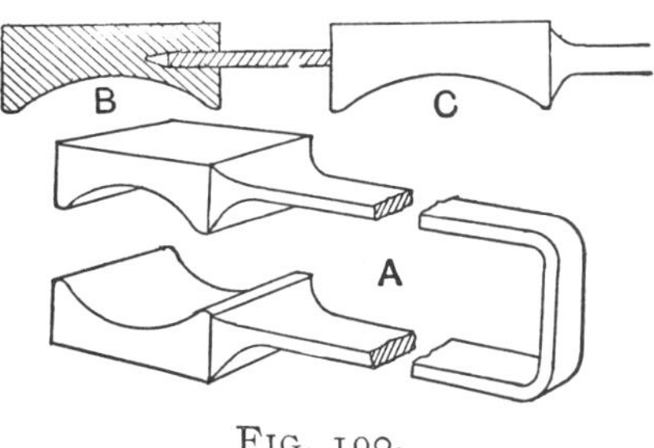

FIG. 190.

by the long spring handle. This spring handle may be made as shown at *B*, by forming it of a separate piece of stock and fastening it to the swage, by making a thin slot in the side of the block with a hot-chisel or punch, forcing the handle into this and closing the metal around it with a few light blows around the hole with the edge of a fuller.

Another method of forming the handle (*C*) is to draw out the same piece from which the blocks are made, hammering down the center of the stock to form the handle, and leaving the ends full size to make the swages.

Swages for large work are made sometimes as shown in Fig. 191. The one shown at *B* is made

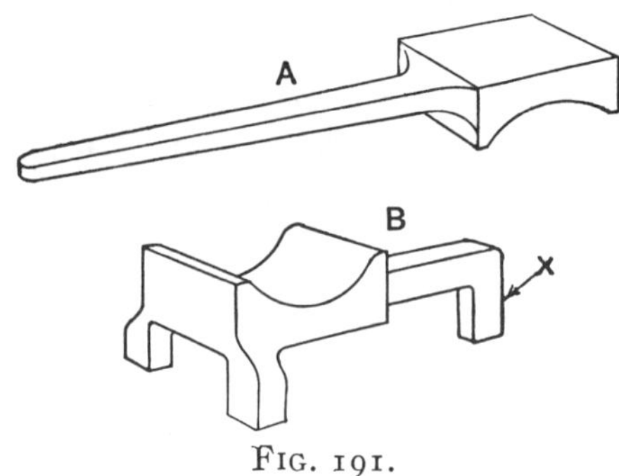

FIG. 191.

for an anvil-die having a square hole, similar to the hardie-hole in an ordinary anvil, near one end. The horn on the swage, at *x*, slips into this hole, while the other two projections fit, one on either side, over the sides of the anvil. These horns, or fingers, prevent the swage from slipping around when in use.

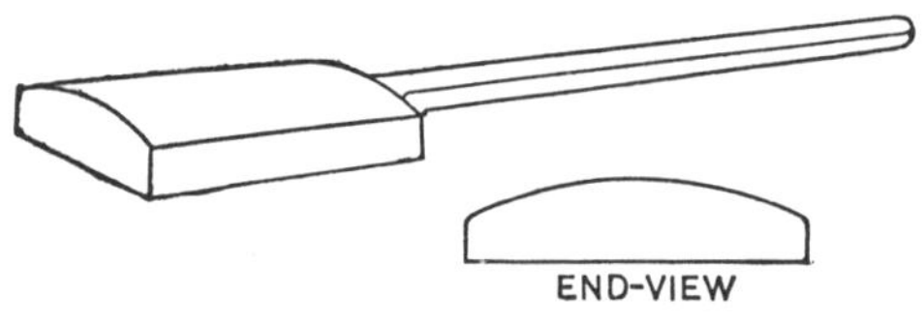

FIG. 192.

Tapering and Fullering Tool.—As the faces of the anvil- and hammer-dies are flat and parallel, it is not possible to finish smoothly between the bare dies, any work having tapering sides.

By using a tool similar to the one shown in Fig. 192 tapering work may be smoothly finished.

Taper Work.—The use of the tool illustrated above is shown in Fig. 193. For roughing out taper work, the tool is used with the curved side

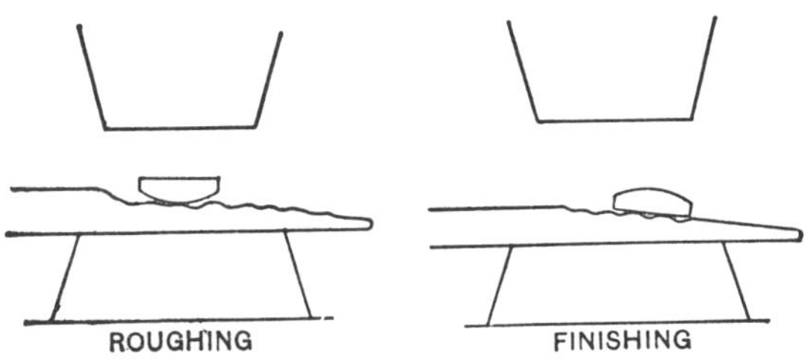

Fig. 193.

down, the straight side being flat with the hammer-die. When finishing the taper, the tool is reversed, the flat side being held at the desired angle and the hammer striking the curved side. This curved side enables the tool to do good work through quite a wide range of angles. If too great an angle is attempted, the tool will be forced from under the hammer by the wedging action.

Fullers.—Fullers such as used for ordinary hand forgings are very seldom employed in steam-hammer work. To take their place simple round bars are used. When much used, the bars should be of tool-steel.

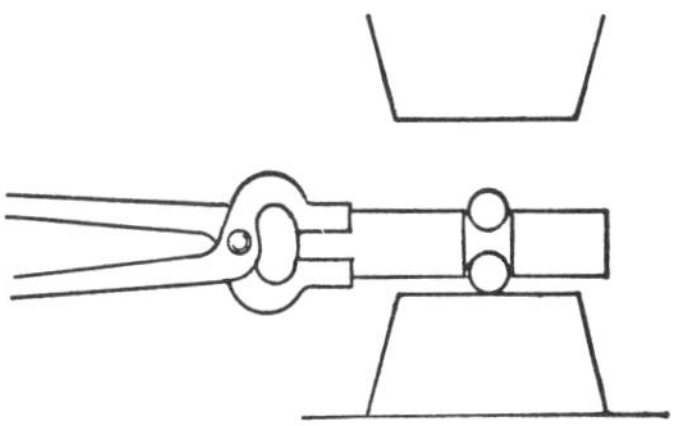

Fig. 194.

One use of round bars, as mentioned above, is illustrated in Fig. 194. Here the work, as shown, has a semicircular groove extending around it, forming a "neck." The groove is formed by placing a short piece of round steel of the proper size on the anvil-die; on this is placed the work, with the spot where the neck is to be formed directly on top of the bar. Exactly above the bar, and parallel to it on top of the work, is held another bar of the same diameter. By striking with the hammer, the bars are driven into the work, forming the groove. The work should be turned frequently to insure a uniform depth of groove on all sides; for, if held in one position, one bar will work in deeper than the other.

Adjusting Work Under the Hammer.—When work is first laid on the anvil the hammer should always be lowered lightly down on it in order to properly "locate" it. This brings the work flat and true with the die-faces; and if held in this position (and care should be taken to see that it is), there will be little chance of the jumping, jarring, and slipping, caused by holding the forging in the wrong position. This is particularly true when using tools, as great care must be taken to see that the hammer strikes them fairly. If the first blow is a heavy one, and the work is not placed exactly right, there is danger of the piece flying from under the hammer and causing a serious accident.

As an illustration of the above, suppose that a piece be carelessly placed on the anvil, as shown in Fig. 195, the piece resting on the edge of the anvil only, not flat on the face, as it should.

When the hammer strikes quickly and hard two

things may happen: either the bar will be bent (as it will if very hot and soft) or it will be knocked into the position shown by the dotted lines. If the hammer be lowered lightly at first, the bar will be pushed down flat, and assumes the dotted position easily, where it may be held for the heavy blows.

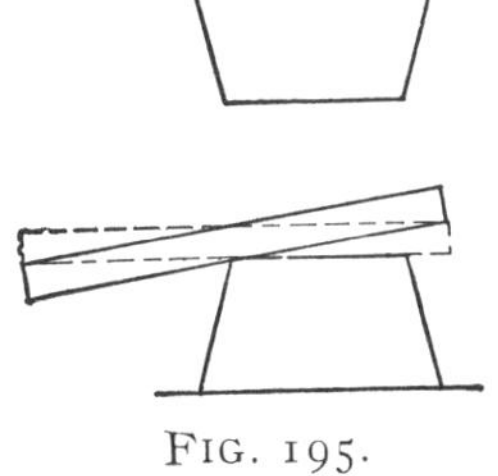

FIG. 195.

Squaring Up Work. — It frequently happens in hammer work, as well as in hand forging, that a piece which should be square in section becomes lopsided and diamond-shaped.

To correct this fault the forging should be held as shown in Fig. 196, with the long diagonal of

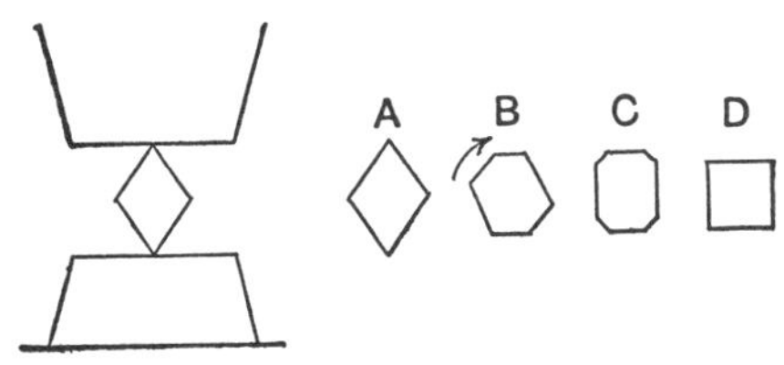

FIG. 196.

the diamond shape perpendicular to the face of the anvil.

A few blows will flatten the work into the shape shown at *B*; the work should then be rolled slightly in the direction of the arrow and the hammering continued, the forging taking the shape of *C*, and, as the rolling and hammering are continued, finally, the square section *D*.

Making Small Tongs.—As an example of manipulation under the hammer, the making of a pair of ordinary flat-jawed tongs is a good illustration.

Fig. 197 shows the different steps from the straight stock to the finished piece.

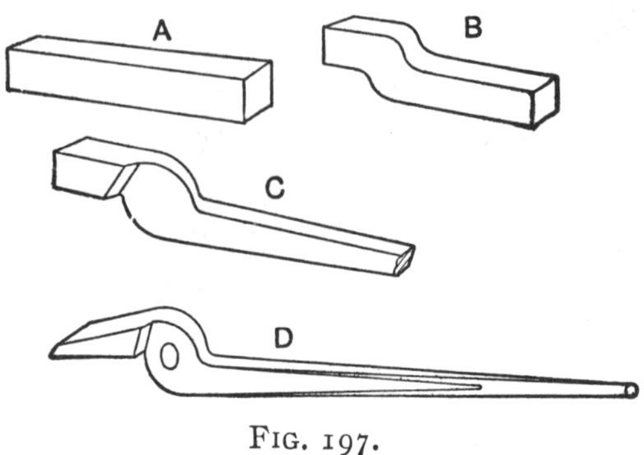

Fig. 197.

The stock is heated to a high heat and bent as shown in Figs. 198 and 199. *A* and *B* (Fig. 198)

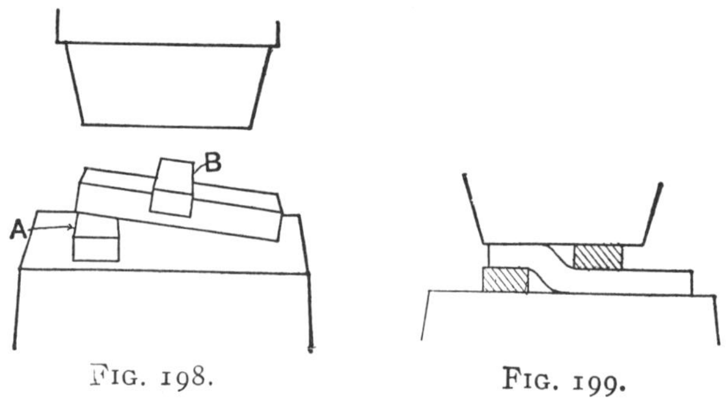

Fig. 198. Fig. 199.

are two pieces of flat iron of the *same thickness.* The stock is placed like Fig. 198, the hammer brought down lightly, to make sure that everything is in the proper position, and then one hard blow bends the stock into shape (Fig. 199).

Fig. 200 shows the method of starting the eye

and working in the shoulder. The bent piece is laid flat on the anvil and a piece of flat steel laid on top, in such a position that one side of the steel will cut into the work and form the shoulder for

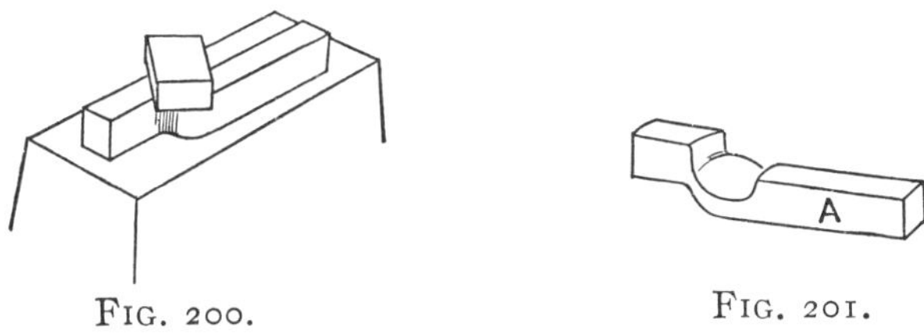

FIG. 200. FIG. 201.

the jaw of the tongs. The steel is pounded into the work until the metal is forged thin enough to form the eye. This leaves the work in the shape shown in Fig. 201. The part *A*, Fig. 201, is afterward drawn out to form the handle, the jaw and eye are formed up, and, lastly, the eye is punched. The forming of the jaw and the punching of the rivet-hole should be done with the hand-hammer, and not under the steam-hammer.

The handle is, of course drawn out under the steam-hammer, but needs no particular description. For careful finishing, the taper tool illustrated in Fig. 192, may be used, or a sledge and swages.

As a general thing, steam-hammer work does not differ very much from forging done on the anvil. The method of operation, in either case, is almost the same; but, when working under the hammer, the work is more quickly done and should be handled more rapidly.

Crank-shafts.—The crank-shaft, shown in Figs.

128 and 129, is a quite common example of steam-hammer work.

The different operations are about the same as described for making it on the anvil.

A specially shaped tool is used to make the cuts each side of the crank cheek. This tool and its use are shown in Fig. 202. When the cuts are

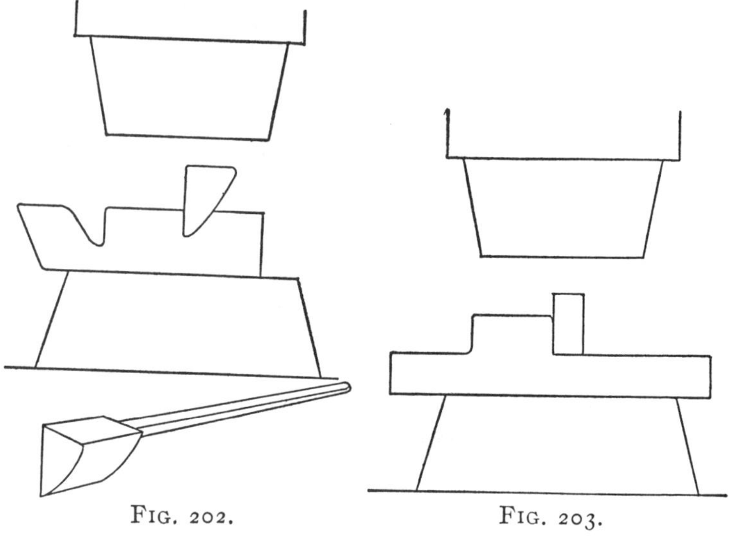

FIG. 202. FIG. 203.

very deep, they should first be made with a hot-chisel and then spread with the spreading tool. If the shoulder is not very high, both operations, of cutting and spreading, may be done at once with the spreading tool.

After marking and opening out the cuts, the same precautions, to avoid cold-shuts, must be taken as are used when doing the same sort of work on the anvil. The work should be held and handled much the same as illustrated in Fig. 131, only in this case the sledge and anvil are replaced by the top and bottom dies of the steam-hammer.

A block of steel may be used for squaring up into the shoulder, as shown in Fig. 203. If a shoulder is to be formed on both sides, one block may be placed below and another above the work, somewhat as shown before in Fig. 194; the round bars in the illustration being replaced with square ones.

Knuckles.—A knuckle such as shown in Fig. 139 would be made by identically the same process as described for making it on the anvil. A few suggestions might be made, however.

After the end of the bar has been split and bent apart, ready for shaping, the work should be handled, under the hammer, as shown in Fig. 204. It

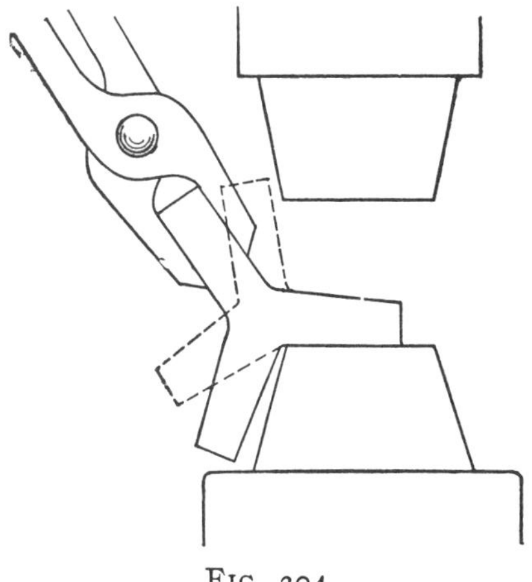

FIG. 204.

should first be placed as shown by the solid lines, and as the hammering proceeds, should be gradually worked over into the position shown by the dotted lines. The other side is worked in the same way.

After drawing out and shaping the ends the knuckle is finished by bending the ends together over a block, in the same way as shown in Fig. 144, the work being done under the hammer.

Connecting-rod. Drawing Out between Shoulders.—The forging illustrated in Fig. 126, while hardly the exact proportions of common connecting-rods, is near enough the proper shape to be a good example of that kind of forging.

The forging, after the proper stock calculation has been made, is started by making the cuts near the two ends, as shown in Fig. 127. The distance, A, must be so calculated, as explained before, that

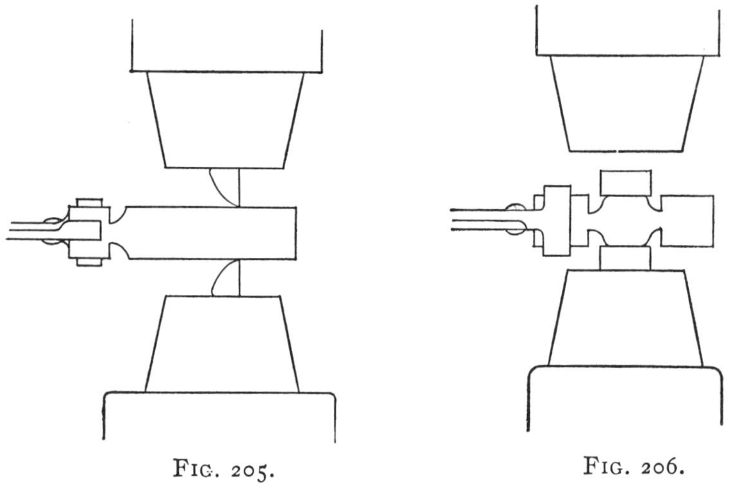

FIG. 205. FIG. 206.

the stock represented by that dimension, when drawn out, will form the shape, 2″ in diameter and 24″ long, connecting the two wide ends.

The cuts are made with the spreading tool used in connection with a short block shaped the same as the tool, or a second tool, the tools being placed one above and one below the work, as shown in Fig. 205.

After making the cuts the stock between them is drawn down to the proper size and finished.

It sometimes happens that the distance *A* is so short that the cuts are closer together than the width of the die-faces, thus making it impossible to draw out the work by using the flat dies. This difficulty may be overcome by using two narrow blocks as shown in Fig. 206.

Weldless Rings—Special Shapes.—It is often necessary to make rings and similar shapes without a weld. The simple process is illustrated in Figs. 155-7. Rings may be made in this way under the steam-hammer much more rapidly than is possible by bending and welding. To illustrate the rapidity with which weldless rings can be made, the author has seen the stock cut from the bar, the ring forged and trued up in one heat. The ring in question was about 10″ outside diameter, the section of stock in rim being about 1″ square. The stock used was about 3″ square, soft steel.

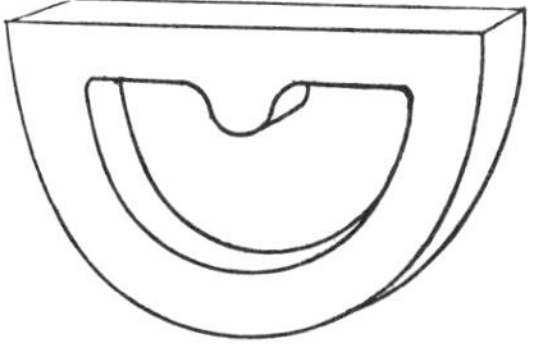

FIG. 207.

A forging for a die to be made of tool-steel is shown in Fig. 207.

This is made in the same general way as weldless rings. The stock is cut, shaped into a disc, punched, and worked over a mandril into the shape shown at *A*, Fig. 208.

The lug, projecting toward the center from the

flat edge of the die, is shaped on a special mandril, the work being done as shown at *B*, the thick side

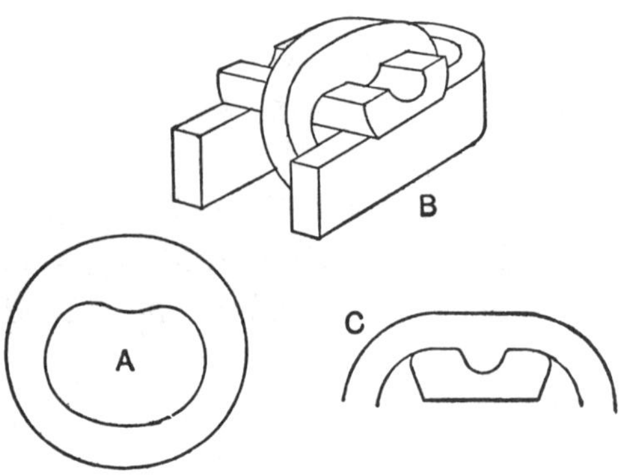

FIG. 208.

of the ring being driven into the groove in the mandril and shaped up as shown at *C*, where the end view of the mandril and ring is shown.

If the flat edge of the die is very long, it may be straightened out by using a flat mandril and work-

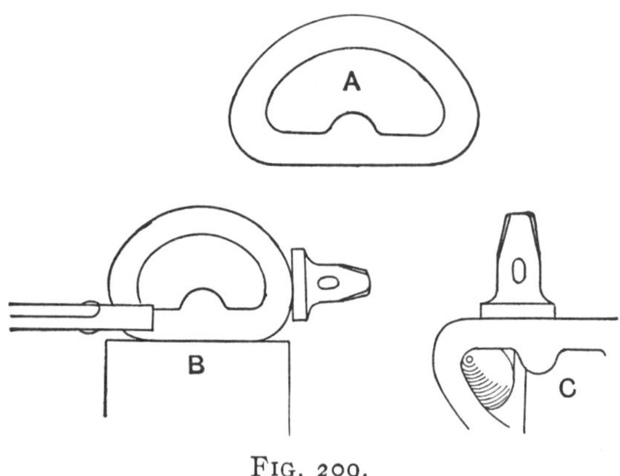

FIG. 209.

ing each side of the projecting lug after the lug has been formed.

The forging leaves the hammer in the shape shown in Fig. 209 at *A*. The finishing of the sharp corner is done on the anvil with hand tools, in much the same way that any corner is squared up, Figs. *B* and *C* giving a general idea of working up the corner by using a flatter.

Punches.—The punches used for this kind of work, and in fact for all punching under the steam-hammer, should be short and thick.

A punch made as shown in Fig. 210 is very satisfactory for general work. This punch is simply a short tapering pin with a shallow groove formed around it about one third of the length from the big end. A bar of small round iron ($\frac{3}{8}''$ is about right for small punches) is heated, wrapped around the punch in the groove and twisted tight, as shown.

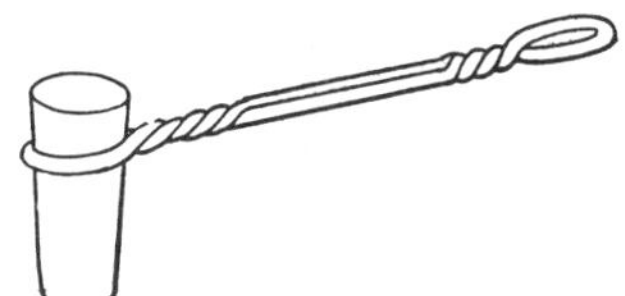

FIG. 210.

The punching is done in exactly the same way as with hand tools; that is, the punch is driven to a depth of about one half or two thirds the thickness of the piece, with the work lying flat on the anvil; the piece is then turned over, the punch started with the work still flat on the anvil, and the hole completed by placing a disc, or some other object with a hole in it, on the anvil; on this the work is placed with the hole in the disc directly under where the punch will come through. The punch is then driven through and the hole completed.

The end of the punch must not be allowed to become red-hot. If the punch is left in contact with the work too long, it will become heated, and, after a few blows, the end will spread out in a mushroom shape and stick in the hole.

To prevent the above, the punch should be lifted out of the hole and cooled between every few blows.

Sometimes, when a hole can be accurately located, an arrangement like that shown in Fig. 211 is used. The punch in this case is only slightly longer than the thickness of the piece to be pierced, and is used with the big end *down* as shown.

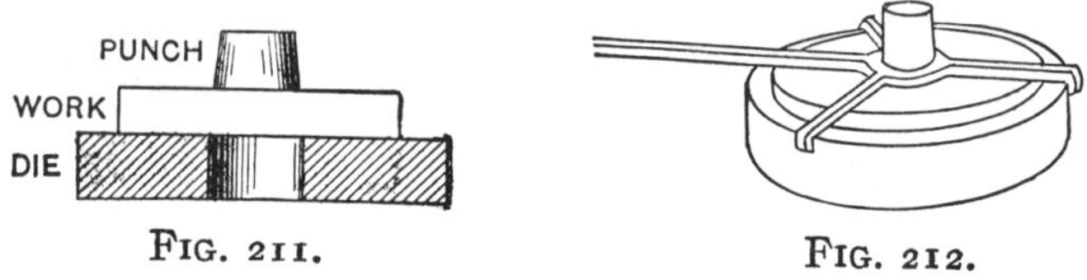

FIG. 211. FIG. 212.

The punch is driven, together with the piece of metal which is cut out, through into the hole in the die, which is just enough larger to give clearance to the punch.

A convenient arrangement for locating the punch centrally with the hole in the die is shown in Fig. 212.

The die should be somewhat larger in diameter than the work to be punched. The work is first placed in the proper position on the die and the punch placed on top. The punch is located by using a spider-shaped arrangement made from thin iron. This spider has a central ring with a hole in the center large enough to slip easily over the punch. Radiating from the ring are four arms, three of which have their ends bent down to fit around the outside of the die, the fourth being longer and used for a handle. The ends of the bent arms are so shaped that where they touch the outside of the die the central hole is exactly over the hole in the die.

After locating the punch with the spider, and while the spider is still in place, a light blow of the hammer starts the punch, after which the spider is lifted off and the punch driven through.

Forming Bosses on Flanges, etc.—A boss, on a flange or other flat piece, such as shown in Fig. 213, may be very easily formed by using a few simple

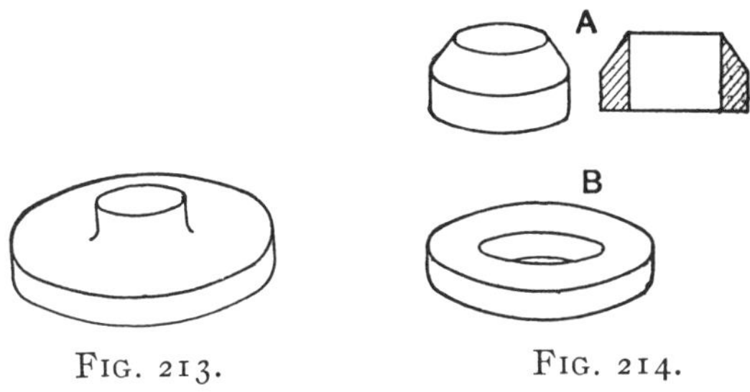

FIG. 213. FIG. 214.

tools. The special tools are shown in Fig. 214, and are: a round cutter used for starting the boss, shown at *A*, which also shows a section of the tool, and a flat disc, shown at *B*, used for flattening and finishing the metal around the boss.

The stock is first forged into shape slightly thicker than the boss is to be finished, as it flattens down somewhat in the forging.

The boss is started by making a cut with the circular cutter, as shown at *A*, Fig. 215, where is also shown a section of the forging after the cut has been made.

The metal outside of the cut is then flattened out, as shown by the dotted lines. This flattening

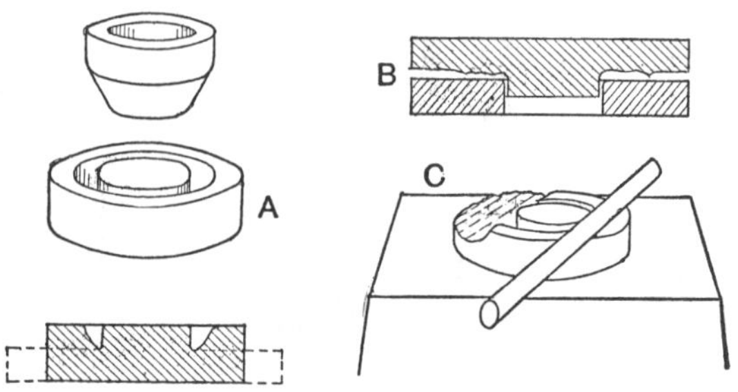

FIG. 215.

and drawing out may be done easily by using a bar of round steel, as shown at *C*. The bar is placed in such a position as to fall just outside of the boss. After striking a blow with the hammer, the bar is moved farther toward the edge of the work and the piece is turned slightly. In this way the stock is roughly thinned out, leaving the boss standing. To finish the work, the forging is turned bottom side up over the disc, with the boss extending down into the hole in the disc, as shown at *B*. With a few blows, the disc is forced up around the boss and finishes the metal off smoothly.

The disc need not necessarily be large enough to extend to the edge of the work; for if a disc as described above is used to finish around the boss, the edge of the work may be drawn down in the usual way under the hammer.

A disc is not absolutely necessary in any case; but the work may be more carefully and quickly finished in this way.

Round Tapering Work.—A round tapering shape, such as shown at *A*, Fig. 216, should be first

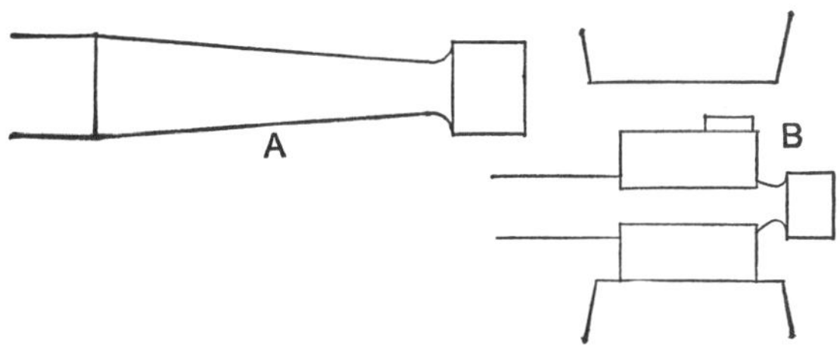

FIG. 216.

roughly forged into shape. It may be started by working in the shoulder next the head with round bars, in the way illustrated before in Fig. 194.

The roughing out may be done with square or flat pieces, using them in much the same way; or one piece only may be used and the work allowed to lie flat on the anvil, with the head projecting over the edge.

After roughing out, the work may be finished with swages. As ordinarily used, the swages would leave the forging straight, with the opposite sides parallel. To form a taper, a thin strip should be held on top of the upper swage close to and parallel with one of the edges, as shown at *B*, Fig. 216. The strip causes the swage to tip and slant, thus forming the work tapering.

CHAPTER VIII.

DUPLICATE WORK.

When several pieces are to be made as nearly alike as possible, the work is generally more easily done by using "dies" or "jigs."

Generally speaking, "dies" are blocks of metal having faces shaped for bending or forming work. The term "jig" may be applied to almost any contrivance used for helping to bend, shape, or form work. As ordinarily used, a jig, generally, is simply a combination of some sort of form or flat plate and one or more clamps and levers for bending.

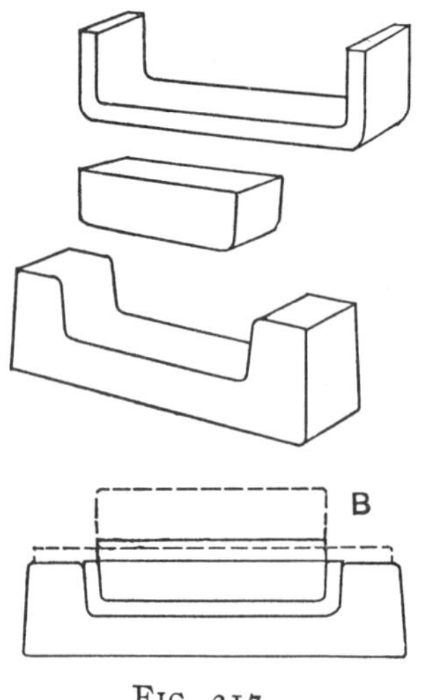

Fig. 217.

Dies, or jigs, for simple bending may be easily and cheaply made of ordinary cast iron; and, for most purposes, left rough, or unfinished.

Simple Bending. — The bend shown in Fig. 217 is a fair example of simple work. The dies for making this bend are two blocks of cast iron made as shown, one simply a rectangular block the size of the inside of the bend to be made, the other a block having on one side a groove the same shape as the outside of the piece to be bent. The blocks should be slightly wider than the stock to be bent.

The stock is cut to the proper length, heated, placed on the hollow block, and the small block placed on top, as shown by the dotted lines at *B*, Fig. 217. The bend is made by driving down the small block with a blow of the hammer.

Work of this kind may be easily done under a steam-hammer; and the dies described here are intended for use in this way, most of them having been designed for, and used under, a 200-lb. hammer.

Dies of this kind may be fitted to the jaws of an ordinary vise, the bending being done by tightening up the screw.

A die such as described above should have a little "clearance"; that is, the opening in the hollow die should be slightly larger at the top than at the bottom. The small, or top, die should be made accordingly, slightly smaller at the bottom.

To make the dies easier to handle, a hole may be drilled and tapped in each block and pieces of round bars threaded and screwed into the holes to form handles. This is more fully described in the following example:

Fig. 218, *A*, is a hook bent from stock $\frac{3}{8}'' \times 1''$, to fit around the flange of an I beam. The hooks were about 6″ long when finished. To bend these, two cast-iron blocks, or dies, were used, shown at

B. The dies were rough castings. Patterns were made by laying out the hook on a piece of 2″ white

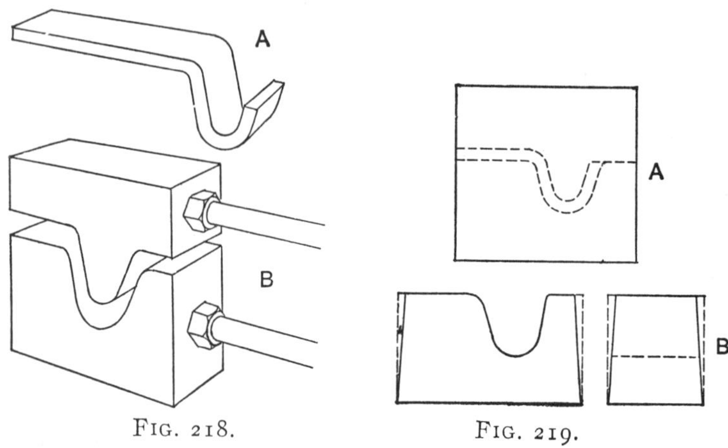

FIG. 218. FIG. 219.

pine and then sawing to shape with a band-saw. The block was "laid off" as shown in Fig. 219, *A*, the sawing being done on the dotted lines. This left the blocks of such a shape that the space between them, when they were brought together with the upper and lower edges parallel, was just equal to the thickness of the stock to be bent.

Patterns of this kind should be given plenty of "draft," which may be quickly and easily done by planing the sides, after the blocks are sawed out, to taper slightly as shown in Fig. 219, *B*, where the dotted lines show the square sides before being planed off for draft as indicated by the solid lines.

When the castings were made, a $^{13}/_{32}$″ hole was drilled in the right-hand end of each block and tapped with ½″ tap. A piece of ½″ round iron about 30″ long was threaded with a die for about 1½″ on each end and bent up to form the handle.

A nut was run on each end and the blocks screwed on and locked by screwing the nut up against them, making the finished dies as shown in Fig. 218. The handle formed a spring, holding the dies far enough apart to allow the iron to be placed between them.

As mentioned before, dies of this kind can be easily made to cover a variety of work, and are very inexpensive. The dies in question, for instance, required about half an hour's pattern work, and about as much time more to fit the handles. Calculating shop time at 50 cents per hour and castings at 5 cents per pound, and allowing for the handle, the entire cost of these dies was less than $1.25.

The same handle can be used for any number of dies of about the same size, and if any one of these dies should break, it can be replaced at a very trifling cost.

Cast-iron dies of this character will bend several hundred pieces and show no signs of giving out, although they *may* snap at the first piece if made of hard iron. On an important job it is generally wise to cast an extra set to have in case the first prove defective.

Almost any simple shape may be bent in this way, and the dies may be used on any ordinary steam-hammer with flat forging faces; and not only that, but, not having to be fastened down in any way, they may be placed under the hammer, or removed, without interfering with other work.

Loop with Bent-in Ends.—For larger work, it is

often better to have a die to replace the lower die of the hammer, as in the case mentioned below.

A number of forgings were wanted like *A*, Fig. 220. The stock was cut to the proper length and

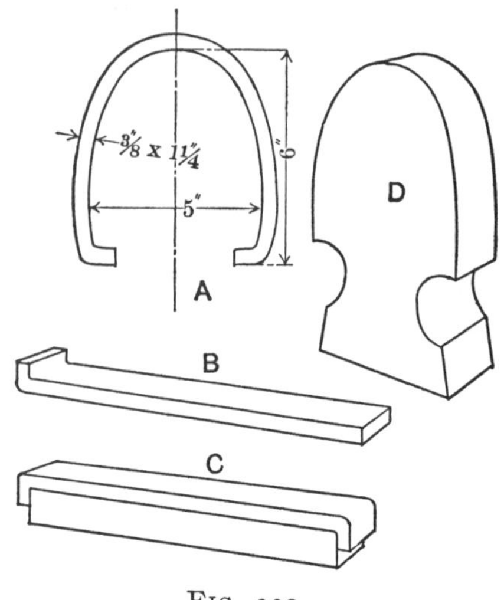

FIG. 220.

the ends bent at right angles. To make all the pieces alike, one end of each piece was first bent, as shown at *B*, in a vise. The other ends of the pieces were then all bent the same way, by hooking the bent end over a bar cut to the proper length and bending down the straight end over the other end of the bar, as shown at *C*. To make the final bend, a cast-iron form was used similar to *D*. This casting was about $2\frac{1}{2}''$ thick, and the dovetail-shaped base fitted the slot in the anvil base of the hammer. When the form was used, the anvil-die was removed and the form put in its place.

The strips to be bent were laid on top of this form and a heavy piece of flat stock, $1'' \times 2''$, bent into a U shape to fit the outside of the forging, placed on top. A light blow of the hammer would force the U-shaped piece down, bending the stock into the proper shape. Fig. 221 shows the operation, the dotted lines indicating the position of the pieces before bringing down the hammer.

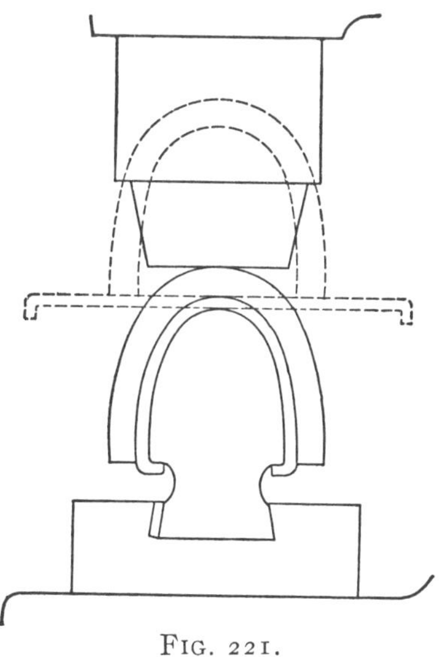

FIG. 221.

The most satisfactory results were obtained by bringing the hammer down lightly on the work, then, by turning on a full head of steam, the ram was forced down comparatively slowly, bending the stock gradually and easily. This was much more satisfactory than a quick, sharp blow.

It is not necessary to have the U-shaped piece of exactly the same shape as the forging. It is sufficient if the lower ends of the U are the proper distance apart. As the strip is bent over the form, it naturally follows the outline; and it is only necessary to force it against the form at the lower points of the sides.

The last bend might have been made by using a second die fastened to the ram of the hammer in place of the U-shaped loop.

Two dies are necessary for much work; but these are more expensive to make. The upper die can be easily made to fit in the dovetail on the ram and be held in place with a key.

Right-angle Bending.—Very convenient tools for bending right angles, in stock ½″ or less in thickness, are shown in Fig. 222. The lower one is made to fit easily over the anvil of the steam-hammer, the projecting lips on either side preventing the die from sliding forward or back. The upper one has a handle screwed in, as described before. Both of these bending tools are made of cast iron, the patterns being simply sawed from a 2″ plank.

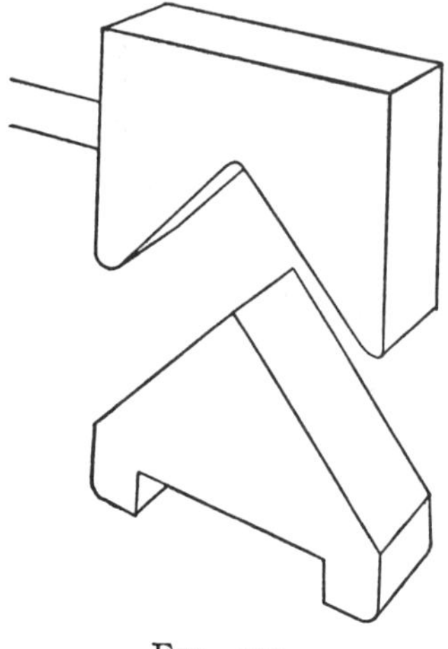

FIG. 222.

Cast-iron dies of this kind should be made of a tough, gray iron, rather than the harder white iron, as they are less liable to break if cast from the former.

Many of the regular hammer dies, that is, the dies with flat faces for general forging, are made of cast iron; but the iron in this case is of another quality—chilled iron—the faces being chilled, or hardened, for a depth of an inch or more.

Circular Bending—Coil Springs.—The dies described before have been for simple bends; the blows, or bending force, coming from one direction only. In the following example, where a complete circle, or more than a circle, is formed, an arrangement of a different nature is required.

The spring shown in Fig. 223 is an example of this kind. In this particular case the bending was done cold; but for hot bending the operation is exactly the same.

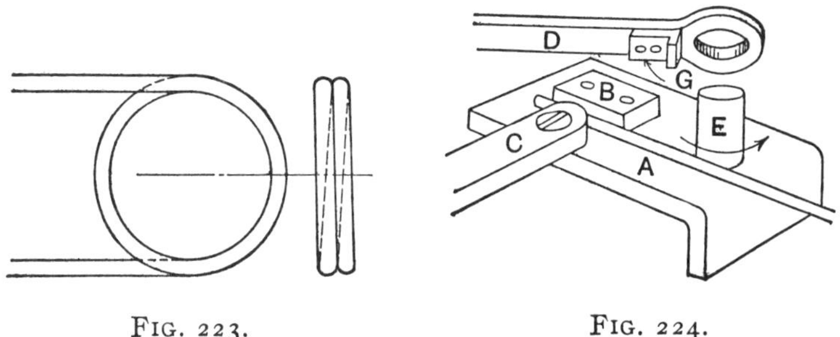

FIG. 223. FIG. 224.

This jig (Fig. 224) was built upon a base-plate, *A*, about $\frac{3}{8}''$ thick, having one end bent down at right angles for clamping in an ordinary vise.

The post *E* was simply a 1″ stud screwed into the plate. *B* was a piece of $\frac{3}{8}'' \times 1''$ stock about 2″ long, fastened down with two rivets, and served as a stop for clamping the stock against while bending. *C* was a lever made of a piece of $\frac{3}{4}'' \times 1''$ stock about 10″ long, having one end ground rounding as shown. This lever turned on the screw *F*, threaded into the base-plate. *D* was the bending lever, having a hole punched and forged in the end large enough to turn easily on the stud *E*. On the under side of this lever was riveted a short piece of iron having one end bent down at right angles.' This piece was so placed that the distance between stud *E* and the inside face of bent end, when the lever was in position for bending, was about $^1/_{64}''$ greater than the thickness of the stock to be bent.

When in operation, the stock to be bent was placed in the position shown in the sketch, the lever *C* pulled over to lock it in place, and the bending lever *D* dropped over it in the position shown. To bend the stock, the lever was pulled around in the direction of the arrow and as many turns taken as were wanted for the spring, or whatever was being bent. By lifting off the bending lever and loosening the clamping lever the piece could be slipped from the stud.

With jigs of any kind a suitable stop should always be provided to place the end of the stock against, in order to insure placing and bending all pieces as nearly as possible alike.

Drop-forgings. — Strictly speaking, drop-forgings are forgings made between dies in a drop-press or forge. Each die has a cavity in its face, so shaped that when the dies are in contact the hole left has the form of the desired forging. One of the dies is fastened to the bed of the drop-press, directly in line with and under the other die, which is keyed to the under side of the drop, a heavy weight running between upright guides. The forging is done by raising the drop and allowing it to fall between the guides of its own weight.

There are generally two or more sets of cavities in the die-faces, one set being used for roughing out, or "breaking down," the stock roughly to shape; another set for finishing.

The dies mentioned above would be known as

the "breaking-down" and "finishing" dies, respectively. Sometimes several intermediate dies are used.

In a general way, the term drop-forging may be used to describe almost any forging formed between shaped dies whether made by a drop-press or other means.

Taking the word in its broadest meaning, the example given below might be called a drop-forging, the work being done between shaped dies.

Eye-bolt — Drop-forging. — The example in question is the eye-bolt given in Fig. 225. The different steps in the making, and the dies used, are shown in Fig. 226.

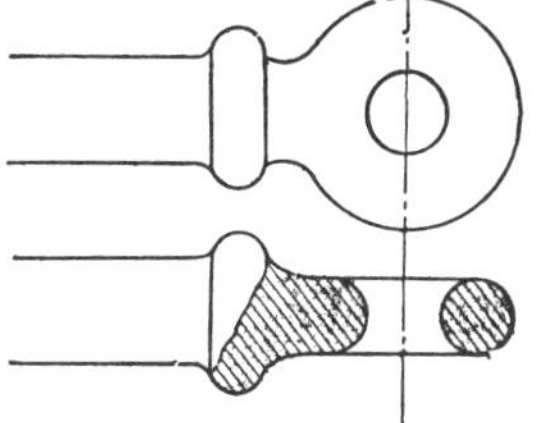

FIG. 225.

Round stock is used, and first shaped like *A*, Fig. 226, the forming being done in the die *B*. This die, as well as the other one, is made in the same way as ordinary steam-hammer swages; that is, simply two blocks of tool-steel fastened together with a spring handle. The inside faces of the blocks are formed to shape the piece as shown.

The stock is revolved through about 90 degrees between each two blows of the steam-hammer, and the hammering continued until the die-faces just touch.

For the second step the ball is flattened to about the thickness of the finished eye between the bare hammer-dies. The hole is then punched, under the hammer, with an ordinary punch.

The forging is finished with a few blows in the finishing die *D*, which is shown by a sectional cut and plan. This die is so shaped that, when the two parts are together, the hole left is exactly the

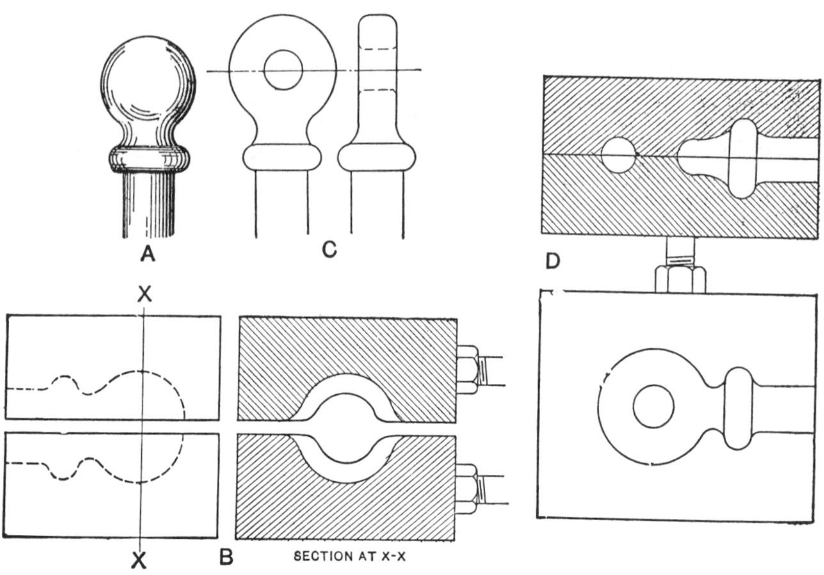

FIG. 226.

shape of the finished forging. In the first die, however, it should be noticed that the holes do not conform exactly to the desired shape of the forging; here the holes, instead of being semicircular, are rounded off considerably at the edges. This is shown more clearly in Fig. 227, *A*, where the dotted lines show the shape of the forging, the solid lines the shape of the die.

The object of the above is this: If the hole is a semicircle in section, the stock, being larger than the small parts of the hole, after a blow, is left like *B*, the metal being forced out between the flat faces of the die and forming 'fins.' When the bar is turned and again hit, these fins are doubled in and make a bad place in the forging.

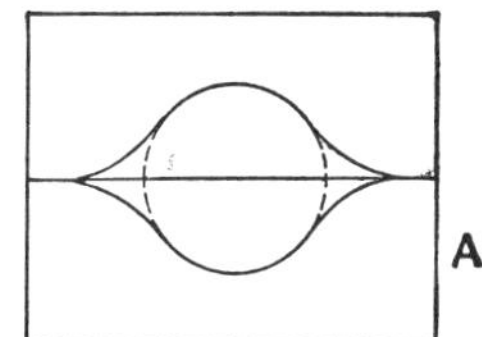

When the hole is a modified semicircle, as described above, the stock will be formed like *C*, and may be turned and worked without injury or danger of cold-shuts.

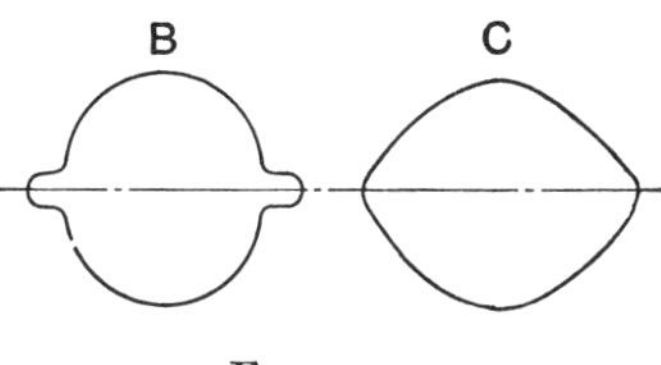

FIG. 227.

Forming Dies Hot.—Making dies for work of the above kind is generally an expensive process, particularly if the work be done in the machine-shop.

Rough dies for this kind of work may be cheaply made in the forge-shop by forming them hot.

The blocks for the dies are forged and prepared, and a blank, or 'master,' forging the same shape and size as the forgings the dies are expected to form is made from tool-steel and hardened.

The die blanks are then heated, the master placed between them, and the dies hammered together, the master being turned frequently during the hammering.

This, of course, leaves a cavity the shape of the master.

When two or more sets of dies are necessary there, of course, must be separate masters for each set of dies. Dies made in this way will have the corners of the cavities rounded off, as the metal is naturally pulled away during the forming, leaving the corners somewhat relieved.

Dies such as described above may be used to advantage under almost any steam-hammer.

For spring hammers, helve hammers, and power hammers generally the die *faces* may be formed the same as above; but the die-blocks should be fastened to the hammer and anvil of the power hammer itself, replacing the ordinary dies.

Cast-iron Dies.—Much drop-forging is done with cast iron dies, and for rough work that is not too heavy they are very satisfactory, and the first cost is very small as compared with the steel dies used for the same purpose.

Drop-forging can be done in this way with the steam-hammer, by keying the dies in the dovetails made for the top and bottom hammer-dies.

Welding in particular is done in this way, as the metal to be worked is in such a soft condition that there is little chance of smashing the die.

CHAPTER IX.

METALLURGY OF IRON AND STEEL.

Classification.—For intelligent working in iron and steel some understanding of their chemical nature and method of manufacture is necessary.

For convenience' sake, the irons and steels ordinarily used in the forge-shop may be divided into three general classes, viz.:

1. Wrought iron.
2. Machine-steel, or low-carbon steel.
3. Tool-steel.

Cast iron should also be considered as being the base product from which the above are derived.

Roughly speaking, the above metals may be considered as mixtures, or, better, compounds of iron and carbon.

There is always present a small percentage of other elements, such as manganese, silicon, sulphur, phosphorus, etc., but for the present these need not be considered.

The percentage of carbon commonly contained in the several materials is about as follows:

Cast iron.	2.50	to	4.50	per cent.
Wrought iron.	.02	"	.50	" "
Machine-steel.	.02	"	.60	" "
Tool-steel.	.70	"	1.50	" "

Cast Iron.—The crude material, from which all iron and steel are manufactured, is iron ore; which is, in its commercial forms, iron oxide. Some of the common ores have much the same appearance and color as iron rust.

To obtain cast iron, the ore is mixed with limestone and melted in a blast-furnace. The blast-furnace is a shell of iron round in section and lined with fire-brick. For a short distance up from the bottom the sides are straight, then rapidly converge, and then contract again to about the same diameter as the bottom. Such a furnace, with its accompanying 'hot-stoves,' is shown, partly in section, in Fig. 228.

The blast-furnace here illustrated is about 80 feet high and 20 feet inside diameter at its largest point.

Heated air, under pressure, is blown into the furnace through water cooled tuyeres placed a short distance above the bed, or bottom, of the furnace.

When the furnace is in operation, there is a bed of burning coke extending somewhat below the level of the tuyeres; on this is a charge, or layer, of mixed ore and limestone, then a layer of fuel, another layer of ore and limestone, etc., until the furnace is nearly filled, more ore and fuel being added as the mass settles down in the furnace. The limestone acts as a flux and helps to carry off, as slag, earthy impurities in the ore. The ore, as it melts is deoxidized; that is, the oxygen is carried off, and the molten iron, being much heavier than the other material in the furnace, sinks to the bottom.

When enough melted iron has collected in the bottom, or hearth, of the furnace, a small hole is opened, and the molten metal flows out and runs

By Courtesy of The Scientific American.

FIG. 228.

into a series of small ditches, much like a gridiron, where it cools, and is then broken up into pieces 4 or 5 feet long. These pieces are called 'pigs'; and in this form cast iron is marketed.

When castings are to be made, these 'pigs' are

remelted in the foundry, in a furnace called a cupola, similar to, but considerably smaller than, the blast-furnace.

It was the custom some years ago to allow the hot gases to escape from the top of the furnace and to blow in cold blast through the tuyeres. Now the top of the furnace is kept closed by means of a cone-shaped casting pulled upward against a conical rim, slanting downward.

When new material is to be added to the charge, the ore or fuel is dumped inside the slanting, funnel-shaped rim and the cone is lowered, allowing the material to slide downward, when the cone is again raised, thus closing the furnace.

Just below the rim a large pipe opens into the furnace. Through this pipe the hot gases are led downward into the 'hot-stoves.'

The hot-stoves are iron cylinders about 20 feet in diameter and 80 feet high, filled with fire-brick having small holes, or flues, extending from top to bottom of the stove. The hot gases rise through one set of flues and descend through another, thus heating the brick to a high temperature.

After leaving the stoves, the gases are carried through underground pipes to a large stack, or chimney, and discharged into the air. When a stove has been thoroughly heated in this way the gases are turned into other stoves, and the cold blast from the blowing-engines is forced through the flues in the heated brick on its way to the blast-furnace.

Each stove is used in turn in this way.

The blast leaving a hot-stove is heated to a temperature considerably over 1000° F. The use of hot blasts effects a considerable saving in the manufacture of cast iron; and its introduction is regarded as marking an epoch in the iron industries.

Wrought Iron.—It will be seen from the table that, while cast iron contains a comparatively large amount of carbon, wrought iron and machine-steel contain very little. It would seem only necessary, then, in order to make either of the two last-named metals, to remove some of the carbon from cast iron. In most cases this is exactly what is done. First, the high-carbon cast iron is made, and then a large part of the carbon is burned out, leaving the low-carbon wrought iron, or machine-steel.

Both wrought iron and machine-steel are made by very similar processes, the essential difference being principally the temperature at which the metals are worked.

Fig. 229 represents a "puddling" furnace used for making wrought iron. The sketch shows a section running the length of the furnace through its center. At *A* is the fireplace; *B*, the hearth, or puddle; and the stack, or flue, at *C*.

A fire is built in the fireplace, and the flames on their way to the stack are deflected downward, by the roof of the furnace, upon the iron lying on the hearth. The iron is thus brought under the influence of the flames without being in direct contact with the fire.

Cast iron, together with hammer scale, or some other oxide of iron, is placed upon the hearth and melted down. The fire is then so regulated as to give an oxidizing flame; that is, more air passed through the fire than can be burned, leaving a surplus of oxygen in the flames which are playing over

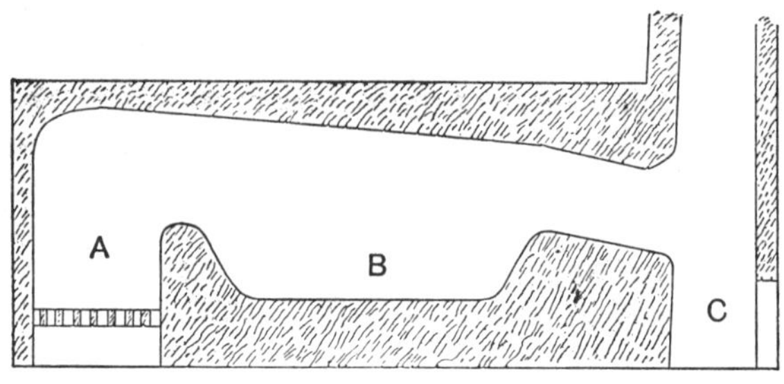

FIG. 229.

the melted iron on the hearth. The oxygen in the flames, as well as that in the hammer scale, or iron ore, melted with the cast iron, gradually burns out the carbon of the cast iron. The melted mass is constantly stirred in order to expose all parts to the influence of the flames.

As a general rule, the more carbon iron contains the easier it melts; so cast iron will melt at a much lower heat than wrought iron. A temperature which is high enough to melt cast iron will leave wrought iron in sort of a pasty condition.

When making wrought iron, as the carbon is burned out of the iron the temperature of the furnace is kept below the melting-point of wrought iron, but above that of cast iron; and, as the carbon is burned out, the metal stiffens and becomes pasty; and, as the process is completed, the pasty mass is worked up into balls, which at the completion of the process are taken from the furnace and hammered or rolled into bars.

There is more or less slag with the iron in the puddle, and some of this slag sticks to the iron, and drops of it are mixed with the iron in the balls. Some of this slag is squeezed from the balls, but part of it remains in small drops all through the mass. When the balls are drawn out into shape, these small drops of slag are lengthened out and form minute streaks running through the length of the bar. These small seams of slag give wrought iron its peculiar characteristics together with its fibrous structure.

Machine-steel.—Machine-steel is variously known as machine-steel, machinery-steel, low-carbon steel, mild steel, and soft steel. The common shop name is machine- or machinery-steel, while the more correct technical term is low-carbon steel.

Machine-steel contains about the same amount of carbon as wrought iron, but does not have the slag seams of the iron. Like wrought iron, it is made by reducing the amount of carbon in cast iron.

Machine-steel may be divided into two classes, open-hearth and Bessemer, both deriving their names from the method of manufacture.

Bessemer steel is made by blowing air through melted cast iron, the oxygen in the air burning out the silicon and carbon, all of the carbon, as nearly as possible, being removed in this way, the proper amount of carbon afterward being added to the steel in the form of "spiegeleisen"—a form of cast iron very rich in manganese, carbon, and silicon.

In this process the cast iron is treated in a vessel known as a converter. The converter is a large barrel-shaped steel or iron vessel, 15 or 20 ft. high, lined with fire-brick, and having a bottom pierced with many small holes through which air is blown. The top is covered, with the exception of a short, spout-shaped opening about 3 ft. in diameter. The converter is mounted on trunnions, and may be turned upside down, right side up, or any intermediate position.

In operation, the converter is turned in a horizontal position, a charge of melted cast iron poured in at the mouth, and the blast turned on. The blast has sufficient pressure to prevent the melted iron from flowing down into the tuyeres in the bottom. The vessel is then turned on its trunnions into an upright position and the air blown through the metal until practically all the carbon has been burned out. The exact condition of the metal is shown by the flame coming from the mouth of the converter, this flame changing in color and volume as the silicon and carbon are gradually burned. When the carbon has been consumed, the converter is again turned on its side, the blast stopped, the necessary amount of spiegeleisen added to give the proper per cent of carbon and manganese, and the contents of the vessel poured into a large casting ladle. From the ladle the metal is poured into moulds and cast into "ingots."

If the "blowing" is continued too long, the iron itself will begin to burn, making the metal "rotten" and crumbly when working.

It seems like a waste of time to burn all the carbon

out and then add more; but it is easier to remove nearly all the carbon and replace some of it than to stop the blow at just the moment when the carbon content is right.

This process derives its name from Sir Henry Bessemer, credited with its invention, and has been used since about 1858.

Open-hearth steel is made by melting together pig iron, cast iron, and scrap iron and steel, and removing the carbon by the action of an oxidizing flame of burning gas.

The process is carried out in a furnace the same in principle as, but more elaborate in construction than, the puddling-furnace used in making wrought iron.

A view and partial section of a large open-hearth furnace is shown in Fig. 230. The fuel used is pro-

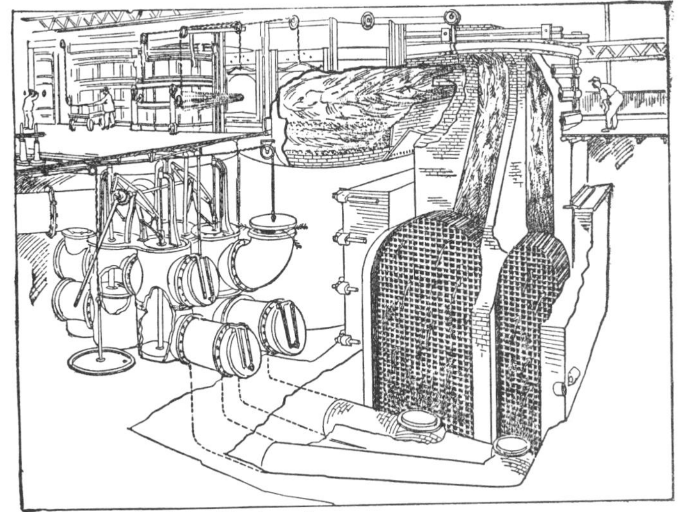

By courtesy of the Scientific American.
FIG. 230.

ducer-gas. This is made by burning coal in air-tight retorts, not enough air being supplied to give complete combustion. The gas from these retorts is forced into the furnace through valves and a mass of heated brickwork pierced with holes. Air is also supplied to the furnace through a similar arrangement, the air and gas entering through openings near each other and combining to make an intensely hot flame. This flame plays over the metal lying on the hearth of the furnace and performs exactly the same work as the flame from the fire in the puddling-furnace.

The hot gases from the furnace, instead of being allowed to escape, are first led through an arrangement of brickwork at the opposite end of the furnace, similar to the heated brickwork through which the entering gas and air were forced.

After a short time the valves, through which the air and gas are admitted, are turned, and the air and gas are forced through the brickwork which has been heating, the flames being then led through the first set of bricks, which is, in its turn, heated, this reversal of the flow taking place several times an hour. This arrangement is very similar to the hot-stoves used with the blast-furnace.

Under the action of the gas-flame the carbon is gradually oxidized, and when it has been reduced to the proper percentage the melted metal is drawn from the hearth and cast into ingots.

This process takes much longer than the Bessemer, but may be stopped at any moment when the steel contains the proper amount of carbon.

As a comparison of the two different processes, it may be said that open-hearth steel, from the nature of the process, may be tested and corrected from time to time while it is being converted, while Bessemer steel is converted so rapidly that none of this testing can be done.

Bessemer steel is used mostly for rails, also to some extent for structural shapes and cheaper boiler-steel, etc.

Open-hearth steel is used for best grades of boiler-plate, structural steel, etc.

The United States Government specifies that sheet steel used in marine boilers must be made by the open-hearth process.

When machine-steel is made, the temperature of the furnace is high enough to keep the metal liquid during the entire process. In this way the slag floats to the top of the iron and does not remain mixed with it, as when making wrought iron.

After sufficient carbon has been taken from the iron, and while the metal is still in a molten condition, it is drawn off from underneath the slag, cast into ingots, and later rolled into bars. This gives the steel a granular, not a fibrous, structure, and leaves it free from the slag contained in wrought iron.

Tool-steel.—Tool-steel may be made by the process outlined above—that is, by making a high-carbon open-hearth or Bessemer steel; but the best steel is made by the "crucible" process.

Ordinary tool-steel contains about 1 per cent carbon, and may be made either by taking some of the carbon from cast iron, which might be done by the methods above, or adding carbon to wrought iron.

The last is the method in common use. In the crucible process, small pieces of wrought iron, steel scrap, and other material rich in carbon are mixed in proper proportions to give the desired percentage of carbon and are placed in a crucible. The crucible is covered with a lid to prevent the oxidation of the melted metal, placed in a furnace and the mixture melted down. When the metal has been melted and properly mixed, the crucible is taken from the furnace and the steel poured into a mold and cast into an ingot, which is afterward rolled into bars.

What was known as "blister steel" was once made in almost the same way "case-hardening" is now done. "Harveyizing" armor plate is also done in somewhat the same way.

The process was based on the fact that when wrought iron is heated in contact with some substance very rich in carbon, it will gradually absorb the carbon from those substances and be converted into high-carbon steel. This is the principle used when making blister steel or in case-hardening.

Steel was at one time commonly made by surrounding bars of wrought iron with charcoal and sealing the bars and the charcoal in air-tight boxes, this being necessary to prevent oxidation during the heating which followed. The boxes were heated to a high temperature and held at that heat for several days. The outside of the bars was carbonized first, thus making a shell, or coating, of tool-steel around a soft, wrought-iron center. In other words, carbon was added to the low-carbon wrought iron and converted it into high-carbon tool-steel. As the heating was continued, the carbon worked in deeper and deeper, but the inside of the bar would not become as highly carbonized as the outside and the steel was "streaky."

After bars were carbonized in this way, they were cut into lengths and welded together, making the steel more uniform in structure, but not nearly as uniform as modern "crucible" steel.

Comparison of Wrought-iron and Machine-steel.— Both wrought-iron and low-carbon steel are chemically about the same; that is, a sample of each may contain about the same amount of carbon, etc., and yet the two materials may be very different.

The broken end of a bar of iron has a stringy, fibrous appearance, while machine-steel shows a more crystalline, grainy fracture. It is this structural condition that marks the distinction between the two metals.

The fiber of the wrought iron is produced by minute, slag seams. Each one of these slag seams is more or less a source of weakness, as the slag, being much weaker than the iron, is liable to give way, causing a crack. The presence of the slag is of some advantage when welding, acting as a flux.

Wrought iron is much more liable to split than machine-steel when being forged, and, while it may be heated and worked at a slightly higher temperature than steel, will not stand as much hammering at a lower temperature.

Machine-steel is stronger than wrought iron, having a tensile strength very much higher.

Machine-steel may be welded easily without a flux; but sound welds are more easily made when a flux is used, the welding being done at a slightly lower heat than the welding heat of wrought iron.

Machine-steel may not be distinguished from wrought iron by the hardening test. Some irons may be slightly hardened, while many low-carbon steels can not be hardened to any appreciable extent. The Government specifications for boiler-plate state that the steel used in boilers shall be capable of being heated to a red heat, plunged in cold water, and then bent double, cold—showing by this that steel does not necessarily harden.

In brief: In a forging, when much welding is to be done, wrought iron has some advantages; but, for general work—particularly when much forging is required—machine-steel is to be preferred, being stronger and less liable to split.

The fibrous structure of wrought iron is well shown by taking a piece 5″ or 6″ long and treating it with weak acid for a day or so. The acid will etch in the iron and leave the fibers of slag standing in relief.

Properties of Wrought Iron, Mild Steel, and Tool-steel.—In brief, the valuable properties of the different metals are as follows:

Wrought iron: Easily welded; easily hammered, or forged, into shape while hot; can be worked to some extent while cold; will not harden to any extent; particularly good for welds.

Mild steel: Easily welded; easily hammered, or forged, into shape while hot; can be worked to some extent while cold; will not harden to any extent; particularly good for forging; stronger than wrought iron.

Tool-steel: Particularly valuable on account of its hardening property; much stronger than mild steel in tensile strength; used principally for making tools and parts of machines where wearing qualities are required; welds with difficulty, sometimes not at all; properties depend to large extent upon percentage of carbon present.

CHAPTER X.

TOOL-STEEL WORK.

Tool-steel.—Ordinary tool-steel is practically a combination of iron and carbon. The kind commonly used for small tools contains about 1 per cent carbon.

Steel which contains a large amount of carbon is known as "high" carbon steel, while that having a small amount is called "low" carbon steel. Steel-*makers* use the word "temper" as referring to the amount of carbon a steel contains; thus a steel-maker speaks of a high-temper steel as meaning a steel containing a large amount of carbon, and a low temper as meaning a small amount of carbon.

Steel is also designated by the number of hundredths of 1-per-cent carbon which it contains. For instance, a one-hundred-carbon steel contains 1 per cent, or one hundred hundredths per cent carbon, a forty-carbon steel contains forty hundredths per cent carbon, etc.

The property of tool-steel which makes it particularly valuable is the fact that it can be hardened to a greater or less degree to suit the purpose for which it is intended.

Hardening.—If a piece of tool-steel be heated

red-hot and then suddenly cooled it becomes very hard. This is known as "hardening." If the reverse be done—the steel heated red-hot and cooled very slowly—it will be softened. This is known as "annealing." In other words—the *speed* at which a piece of steel is cooled from a high heat determines its hardness; thus, if steel is cooled *very fast* it becomes very hard; if cooled *very slowly* it is softened, and by varying the speed of the cooling the hardness of the steel may be varied.

The proper heat from which the steel should be cooled varies with the percentage of carbon in the steel. As a general rule, the greater the amount of carbon, the lower the hardening heat—that is, a "high" steel will harden at a much lower temperature than a "low" carbon steel.

The only way to determine the proper heat at which to harden any particular kind of steel is by experiment. This may be easily done as follows: A small piece of the same kind of steel as that to be hardened is hammered out into a bar about $\frac{1}{4}''$ or $\frac{3}{8}''$ square. The end of this bar is heated until it shows dull red and then cooled in cold water. The end should be tried with a file and about $\frac{1}{4}''$ broken off over the corner of the anvil.

It will probably be found that the steel may be filed, and breaks with difficulty, the grain of the broken end being rather coarse and somewhat stringy. The same experiment should be repeated at a slightly higher heat. This time the steel will be harder—shown by its being harder to file and more easily broken—and the grain will be slightly finer. This experiment should be repeated, raising the heat slightly each time, until a heat is reached which, after cooling, leaves the steel so hard that the file will slip over without catching at all, and so brittle that it snaps very easily. When broken, the break shows a very fine, even grain. This is the proper heat at which to harden that particular kind of steel, and is called the "hardening heat."

If the experimenting be continued it will be seen that each additional increase of temperature, *above the hardening heat*, increases the coarseness of the grain and makes the steel very brittle, indicating that the steel, when hardened at these higher heats, grows coarser and less fine in texture, and, consequently, is not as strong, and will not hold as good a cutting edge, as if hardened at the proper "hardening" heat.

The proper heat at which to harden any kind of steel is, as noted above, that particular heat that gives the steel the finest grain and leaves it file hard.

The two general laws of hardening are these:

1. The more carbon a steel contains the lower the heat at which it may be properly hardened.

2. The faster steel is cooled from the hardening heat the harder it becomes.

Tempering.—Giving a piece of steel or a tool the proper degree of hardness to do the work for which it is intended is known as "tempering."

The steel-workers use the word "temper" in a very different sense from the steel-makers. "Temper" is used by the tool-maker, or tool-smith, as meaning the hardness of a tool or piece of tempered steel, regardless of the amount of carbon it contains.

Tools hardened as described above (heated to hardening heat and cooled in cold water) are too hard and brittle for most uses, and must be softened somewhat to fit them to perform the work they are intended for.

The operation of slightly softening the hardened steel is known as "drawing the temper," and this is accomplished by slightly reheating the previously hardened steel.

If a piece of hardened steel be heated to a temperature of about 430° F. it will be very slightly softened and toughened, being left about hard enough for engraving-tools, small lathe-tools, scrapers, etc. If the heat be raised to 460° F. or 500° F., the hardness is about right for taps, dies, drills, etc. Reheating to 550° F. or 560° F. makes the hardened steel about right for cold-chisels, saws, etc.; while a temperature of 570° F. leaves very little hardness in the steel—just about right for springs. When a temperature of 650° F. is reached the "temper" is all gone and the steel is left soft enough to be easily filed. With a slight increase of temperature above this point the steel becomes red-hot.

These temperatures to which the steel is reheated may be measured in several ways. One way would be to heat a bath of oil to the proper temperature and, after hardening, dip the tools in this until they were of the same temperature as the bath. This would answer for tempering on a large scale, but is hardly practical when only a few tools are to be treated.

The steel itself furnishes about the easiest means of roughly determining this temperature. If a piece of steel or iron be polished bright and heated, a thin scale forms on the outside, which changes color as the temperature is increased. When the scale first commences to appear, at a temperature of about 430° F., the surface of the steel seems to turn a very pale yellow; as the temperature increases and the scale grows thicker, this yellow becomes darker, changes into brown, which becomes tinged with red, turning into light purple, dark purple, and finally blue. These colors are due to the thin oxide or scale formed and show nothing except the temperature to which the metal was last heated.

A piece of wrought iron or soft steel will, when heated, show these colors as well as tool-steel. The colors are permanent and remain after the metal is cooled. The colored scale is very thin and may be easily removed by polishing.

If the tool is not properly hardened in the first place the fact that it shows the proper temper color on the outside means *nothing*.

About the only way to test the temper of a tool is to try it with a file, and even then the grain may be too coarse, due to hardening at too high a heat.

The complete process of tempering a tool (i.e., giving it the proper degree of hardness to perform its work) consists of first hardening, by cooling suddenly from the hardening heat, and then slightly softening, by reheating to a comparatively low temperature.

After the reheating the steel may be suddenly cooled or left to cool in the air. Sudden cooling leaves it slightly harder.

The higher the temperature of the reheating, up to a visible red heat, the softer the steel and the "lower" the temper.

If the steel is by accident or otherwise reheated to too high a temperature when drawing the temper, the tool must be rehardened and the temper again drawn.

When there is any doubt as to the proper heat at which to harden any piece of steel it is much better to harden at too low rather than too high a heat. If hardened at too low a heat it may be reheated and again hardened at a proper heat, but if too high a heat is used the first time there is no way of detecting the fact, and the tool will probably break the first time used.

As a general rule, a tool hardened at too high a heat will have a crumbly and scratchy cutting edge.

Tempering Tools.—In practice tools may be divided for convenience in tempering into two general classes:

First, tools which have only a cutting edge tempered, such as most lathe-tools, cold-chisels, etc.

Second, tools tempered to a uniform hardness throughout, or for a considerable length, such as dies, reamers, taps, milling-cutters, etc.

Tempering Tools when Only an Edge is Hardened —Cold-chisel.—The method of tempering a cold-chisel will serve as an example of the tempering of tools in the first class, the only difference in the tempering of various tools in this class being the temperature to which the tools are reheated, as shown by the "temper color."

A table showing the temperatures to which various tools should be reheated after hardening to properly "draw the temper," together with the so-called "temper colors," or color of scale, corresponding to these temperatures, is given on page 248.

After the chisel has been forged it should be allowed to cool until black. The cutting end is then heated to the proper hardening heat for two or three inches back from the edge, care being taken not to heat the steel *above* the hardening heat. (Steel should always be hardened at a *rising* heat.) The heated end is then hardened by cooling about two inches of the point in cold water, the end being left in the water just long enough to cool it. The chisel is then withdrawn from the water and the end polished with a piece of emery-paper, old grindstone, or something of that character.

As part of the chisel is still red-hot the heat from this hot part will gradually reheat the cold end, thus "drawing the temper." "Temper colors" will begin to show next the heated part, and as the cold end is reheated the band of colors will move toward the point of the chisel. When a deep bluish-purple color shows at the cutting edge the tool is again cooled in order to prevent further reheating and softening of the steel.

Should part of the chisel be still red-hot when the end is cooled the second time, then only the end should be dipped in the water and the tool held there until all of the chisel is black, when the entire length may be cooled.

If it were a lathe-tool being tempered the process would be the same excepting the tool should be cooled the second time when the *yellow* scale appears at the cutting edge.

When hardening the ends of tools as described above, the tool while in the water should be kept in constant motion to prevent cooling the steel along a sharp line, as well as to keep up a circulation of water around the cooling metal.

Tempering Tools of the Second Class (Hardened all Through).—Tools of the second class (those of uniform hardness throughout) may be tempered as follows: The whole tool is first heated to a uniform hardening heat and cooled completely, thus hardening it throughout. The surface is then polished bright and the temper drawn by laying the tool on a piece of red-hot iron until the surface shows the desired color, generally dark yellow or light brown for such tools as taps, dies, milling-cutters, etc.

While reheating on the iron the tool should be turned almost constantly, otherwise the parts in contact with the iron will become overheated, and consequently too soft, before the other parts are hot enough.

Sometimes the reheating is done on a bath of

melted lead or heated sand. Large pieces are sometimes "drawn" over a slow fire or on a sheet of iron laid over the fire.

Recalescence.—There is a peculiar fact which helps to determine the proper hardening temperature of a piece of steel. If a piece of steel be heated to about a bright-red heat and allowed to cool, it will cool gradually until a temperature is reached at which it seems to grow hotter; that is, it grows darker in color, and then, when the critical temperature is reached, it becomes lighter for an instant and then gradually cools down. The temperature at which this seeming reheating takes place is about the proper hardening heat.

This phenomenon is known as "Recalescence." An attempt is made in Fig. 231 to illustrate the

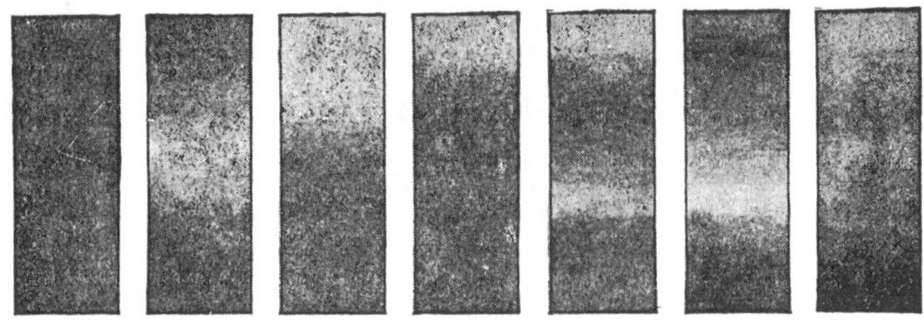

FIG. 231.

action of the heated bar at the point of recalescence. *A* shows the heated bar as it comes from the fire—the hottest part showing lightest. At *B* the steel has cooled slightly and the heat of recalescence begins to show at the light point about the centre of the bar. At *C* the first streak has shifted somewhat and the end begins apparently to reheat, this second streak gradually moving up as illustrated at *D*, *E*, and *F*, until at *G* the bar has passed the critical temperature and cools down normally.

This illustration is somewhat the same as that shown by Howe in his "Metallurgy of Steel," which gives an excellent explanation of this phenomenon.

The temperature at which this reheating occurs depends upon the amount of carbon in the steel, being higher for the lower carbon steel, and it is at about this temperature that the steel will harden properly.

The hardening heat of steel is often described as a "cherry-red" heat. This term is very misleading and means very little. Such an authority as Metcalf says: "Cherries are all shades from very light yellow to almost black; and 'cherry' heat seems to mean almost any of these various colors."

It is a good plan when taking the steel from the fire to hold it for an instant in the shadow of the forge, as the hardening heat may be distinguished with more certainty in this way. The color of the heat will appear quite different here than in the sunlight, and there is a better chance of obtaining uniform heat by judging in the shadow than in the open sunlight, which varies so much in intensity.

Rate of Cooling—Different Hardening Baths.—The more quickly steel is cooled for a hardening heat (everything else being equal), the harder and more brittle the steel is made.

Files, wanted very hard, are hardened by cooling in a bath of cold brine; as the brine cools the steel faster than water the steel is left harder than if hardened in water.

Springs wanted tough and not very hard are cooled in oil, as oil cools much slower than water.

Sometimes articles delicately shaped and liable to crack when hardening are cooled in water having a thin film of oil on top; the oil sticks to the steel as it is plunged into the water and the steel is not cooled quite as quickly as in pure water.

The faster steel is cooled the more danger there is of cracking.

Heating and Cooling—Importance of Uniform Heating.—The greatest care must be taken when hardening to have the steel *uniformly* heated. It must not be left in the fire one minute longer than is necessary to accomplish this; but it *must be uniformly heated* or the results are liable to be disastrous. Take a milling-cutter, for example, with sharp-pointed, projecting teeth. The points of these teeth may become much hotter than the body of the cutter while being heated. If dipped while the points are hotter, they are almost certain to crack off.

Too much importance can not be attached to uniform heating. It is safe to say that probably the failure of three-quarters of the work spoiled in hardening is caused by improper heating.

When it is necessary to heat milling-cutters and

flat tools in an open fire, it is a good plan to lay a piece of thin, flat iron on the fire and heat the steel on this. The steel does not then come in direct contact with the fire and may be more uniformly heated.

For heating taps, small-end mills, etc., a piece of pipe may be laid through the fire and the tools heated in this. The pipe forms a crude muffle, which is very satisfactory for such work.

The most satisfactory way to harden is to use a gas-furnace, but this is not always obtainable.

Lead Hardening and Tempering.—A bath of lead is frequently used for heating both when hardening and drawing the temper.

When hardening the lead is heated red-hot (hardening heat of the steel) and the tools to be hardened are held in the lead until heated to the proper temperature.

The top of the hot lead is kept covered with charcoal to prevent oxidation, otherwise, the lead when exposed to the air would be rapidly oxidized and wasted.

The steel is cooled in the ordinary way.

This is a very satisfactory way to harden, as the steel may be very uniformly heated. For small work the lead may be heated in an ordinary ladle; for larger pieces some special arrangement is necessary.

When drawing the temper the lead is not heated as hot as when hardening, and the pieces to be tempered are laid on top of the melted lead. The steel, being lighter than lead, will float on top and may then be easily watched during the heating. The pieces to be tempered are polished and heated until the proper colors appear—the same as when heated in the ordinary way.

Warping in Cooling.—When heated steel is cooled it contracts, and unless contracting takes place uniformly on all sides the piece is liable to be warped, or sprung, out of shape. If, for instance, a long, thin, flat piece of steel were to be hardened by dipping into the cooling bath edgewise or flatwise it would probably spring out of shape. If dipped endwise the piece would be cooled from all sides at once and would stand a better chance of coming out of the cooling bath straight.

As a general rule, it is better to dip cylindrical and long thin pieces endwise; round, thin discs and square flat pieces edgewise.

Cooling Thin Flat Work.—Very thin flat work of uniform thickness is easily hardened as follows: The piece is heated to a hardening heat and cooled between two heavy plates of iron having flat faces smeared with oil. The piece is laid on one plate and the other quickly laid on top. This leaves the work hard and very true and flat. Pieces which would be warped all out of shape if cooled in water or oil may be easily hardened in this way. The temper is drawn in the ordinary way.

Hardening Files—Straightening Long Thin Work.—The hardening of files is a good example of the treatment of long thin work; and the method employed may be used to advantage for many other pieces.

The files are heated in a pot of red-hot lead. They are placed in this pot on end, and when properly heated are plunged end first (being held in a vertical position) into a vat of brine.

The files nearly always warp somewhat when hardened, and when the warping is slight are straightened as follows: Across the top of the brine vat are fastened two wooden strips about two inches apart, joined by two iron pins about six inches from each other. The hardener draws his file from the brine before it is entirely cold. The metal has just heat enough left to cause the water on the surface of the steel to disappear almost instantly. The file is then placed between the pins, under one and over the other, with the concave side up, as shown in Fig. 232, which shows one of

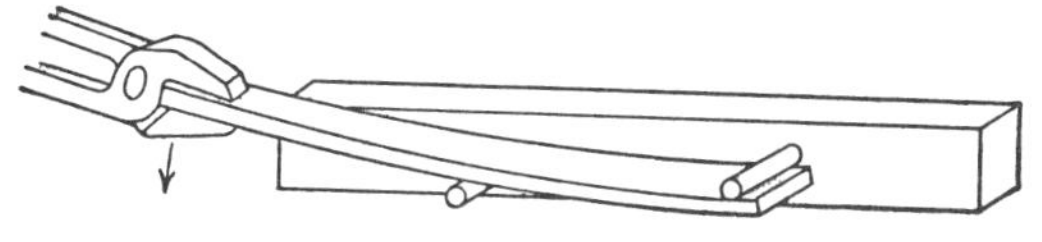

FIG. 232.

the side strips removed. The hardener then bears down on the end of the file, springing it straight, and at the same time pours some of the cold brine on top of the concave part. This will generally straighten out the file and leave it perfectly true. Of course if the files are too badly warped there is nothing to do but reheat, straighten, and harden again.

Bad Shape to Harden.—There are some shapes

which are very difficult to harden. Fig. 233 shows a sectional view of a steel bearing which should be hardened very hard. The body of the bearing is thick and contains proportionately a large volume of metal, while the flange is very thin and light and joins the body in a sharp angle, making a bad shape to harden. The thin flange cools almost instantly as it strikes the water, while the body takes some seconds to cool and by that time the flange is set. As the body contracts in cooling it pulls away from the flange, cracking in the sharp corner.

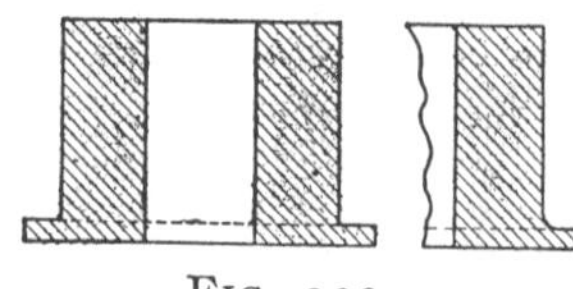

FIG. 233.

Of course shapes like this will not *always* crack, but there is always a strong tendency to do so when a thin body of metal joins a thick one with a sharp corner between the two parts. This danger can be lessened by leaving a fillet in the corner as shown in the side sketch. This equalizes the strain somewhat by not leaving a distinct line between the thick and thin parts.

Milling-cutter teeth when made with a sharp angle at the bottom are liable to crack in hardening. If left with a slight fillet between them they very rarely crack if properly heated.

Tempering Springs—Blazing Off.—Spring tempering done in oil is a good example of work where the temperature of the reheating is determined independently of the "temper colors."

This method of tempering springs is known as "blazing," and gives about as reliable results as any, on a small scale, for ordinary work. The spring is heated to a hardening heat and cooled in oil. To draw the temper, the spring, still wet with the oil, is reheated (ordinarily in the blaze of the forge) until the oil blazes up and then plunged for an instant into the oil-bath and again reheated until it blazes. This is continued until the oil blazes uniformly over the entire spring at the same time.

Springs are generally not uniform in thickness, and the thin parts heat more quickly than the thicker. This momentary plunge into the oil cools the thin parts somewhat and affects very little the thicker parts. As the reheating is continued all parts of the spring are thus brought to the same temperature at the same time. If the reheating were continued without this partial cooling, by the time the thicker parts were hot enough to have the proper temper, the thinner parts would be heated to too high a temperature and have no temper left.

Animal oil, and *not* mineral oil, should be used for this kind of work, as the mineral oil is too uncertain in its composition and will sometimes blaze at one temperature, sometimes at another, while the animal oil is fairly uniform in its composition and generally blazes at about the same temperature. Lard- or fish-oil is a good material for this purpose.

Another way of tempering springs (which is rather risky but which is sometimes used) is to harden the spring in the ordinary way in water. It is then reheated over the fire, and to test the temperature from time to time a dry, pine splinter is scraped over the edge of the spring. As soon as the minute shavings thus made will catch fire the right temperature is supposed to be reached, the burning of the wood in this case taking the place of the burning oil mentioned above.

Sometimes when no pine splinters are convenient even the hammer-handle is made to serve the purpose.

These last are not processes to be recommended, but are given as illustrations of how the temperature of reheating for drawing the temper may be determined in a variety of ways, viz., by the blue color of the scale; by the blazing of the oil; by the burning of the wood.

Hardening to Leave Soft Center.—Another method of tempering used for milling-cutters and taps which has proved very satisfactory is as follows:

The tools are heated in the ordinary way and are cooled in water, but are not left in the water long enough to become completely cold, being drawn out of the water as soon as the "singing" stops. (When red-hot metal strikes water the water in immediate contact with the metal starts to boil, and this boiling produces a decided humming or singing noise and a throbbing sensation easily felt through the tongs. This ceases when the outside of the metal cools to about the temperature of boiling water.)

When the tool is drawn out of the water it is *instantly* plunged into lard-oil and left there for a very short time (depending upon the size of the tool) and then withdrawn. It is then held in the flame of the forge or near the fire until the oil on the outside just commences to smoke, when it is again plunged for an instant into the oil and again reheated, this being continued until the oil smokes evenly all over the tool, when the tempering is complete and the tool may be cooled off.

The object of the method is this: The first cooling in water hardens the outside and cutting edges of the tool. The tool is then taken from the water and plunged into oil while the inside is still comparatively hot. As the oil conducts the heat more slowly than water, the cooling of the tool is continued in the oil, thus leaving the center rather tougher than if hardened in water. But even here the metal is not completely cooled, but taken from the oil-bath while there is still some heat left in the center. This heat in the center helps to draw the temper of the outside, and consequently the tool is reheated much quicker than if entirely cooled. The smoking oil serves to indicate the proper temperature to which the reheating should be carried.

With a little practice the tool can be withdrawn from the oil-bath while there is still heat enough left in the central part to draw the temper. In this way no reheating in the fire is necessary, the tool being simply taken from the oil, allowed to reheat itself until the oil commences to smoke, and then plunged in water to prevent further reheating.

Annealing.—It may be said that annealing is the reverse of hardening. To go back to first principles: If a piece of steel be heated to a proper "hardening heat" and cooled very suddenly, the steel is left very hard. The faster it is cooled the harder the steel. On the other hand, if the steel be cooled from this hardening heat very slowly it is left soft, and the slower it is cooled the softer it becomes. This softening process of heating and cooling slowly is known as *annealing*, and steel so treated is called *annealed steel*.

Water Annealing.—The quickest way of annealing is what is known as "water annealing." In doing this the steel is heated until it just shows very dull red when held in a dark place, and is then cooled in water. This method leaves the steel soft enough to be worked, but not as soft as it would be if heated to a hardening heat and cooled very slowly.

A bar of steel if hammered until cooled below a red heat is rather hard, and can be made somewhat softer and put in a better condition to work by water annealing.

Water annealing is quite often used for work of the following nature: A drill or tap is sometimes broken off in a piece of work and must be softened before it can be removed. Such work is generally wanted in a hurry, and water annealing is resorted to to soften the broken piece.

Soapy water gives good results for water annealing.

Annealing at the Forge.—A common way of annealing is to bury the heated steel in the cinders on the forge and keep it there until it is cold. This method is very satisfactory for ordinary work. Still another way, and about the most satisfactory, is to bury the metal in a box filled with common lime. The steel cools slowly in this, and is left in very good shape for working.

The object in each case is to cool the metal as slowly as possible by keeping the air from it, as heat is lost to the air very rapidly. When the steel is buried in some material which does not conduct heat readily, it of course cools very slowly and is left that much softer.

Box Annealing.—Sometimes a piece of polished steel must be annealed without raising any scale on the surface, such as would be left by any of the methods described before. To do this the air must be kept from the metal, both while it is being heated and while cooling. The steel can be buried in an iron box, filled with ground bone, burned leather, or other carbonaceous materials and sealed air-tight. The box may be slowly heated to the right temperature and allowed to cool very slowly, the steel being removed from the box after it is cold.

There is a patented process for annealing polished steel which is said to leave the metal as bright and polished as it was before annealing. The method is about as follows: The pieces to be annealed are placed in a piece of large pipe having a cap on one end; into this cap is screwed a small gas-pipe which extends back through to the outside of the furnace. When the pieces are all in the large pipe, a second cap is screwed on the open end. This cap has a small hole drilled in it. While the large pipe containing the steel is being heated gas is run into it through the small gas-pipe. This gas fills the pipe and escapes and burns at the small hole through the second cap.

Ordinary illuminating-gas is used. Oxygen is necessary to form scale on the steel, and as the gas (containing very little or no oxygen) which fills the pipe drives out the air, there is no oxygen left to form scale and discolor the steel, consequently the steel comes out of the pipe as bright as when it went in. The pipe, of course, is kept full of gas until the steel is cold.

Mr. William Metcalf, in his book on Steel, gives as a substitute for the above-patented process (known as the Jones process) the following method (this description is taken verbatim from his book): "Let a pipe be made like a Jones pipe without a hole in the cap or a gas-pipe in the end. To charge it first throw a handful of resin into the bottom of the pipe and screw on the cap. The cap is a loose fit. Now roll the whole into the furnace; the resin will be volatilized at once, fill the pipe with carbon or hydrocarbon gases, and with the air long before the steel is hot enough to be attacked.

The gas will cause an outward pressure, and may be seen burning as it leaks through the joint at the cap. This prevents air from coming in contact with the steel. This method is as efficient as the Jones plan as far as perfect heating and easy management are concerned. It reduces the scale on the surfaces of the pieces, leaving them a dark-gray color and covered with fine carbon or soot. For annealing blocks it is handier and cheaper than the Jones plan, but it will not do for polishing surfaces."

File blanks (the shaped pieces of stock ready to have the teeth cut) are annealed by packing them in cast-iron boxes 3½″ or 4″ long, 1″ deep, and 8″ or 10″ wide, with just a little sprinkling of some carbonaceous material over the steel. The box is closed by an iron cover which fits inside the box and comes about an inch and a half below the top of the sides.

The box is made practically air-tight by packing fire-clay (the damp dust or grit which collects beneath the grindstones is sometimes used) around the inside the box on top of the cover.

These boxes are placed in a furnace and heated for about forty-eight hours and then drawn out and covered with sheet-iron covers lined with asbestos, where they cool very slowly. A box put under a cover Saturday is expected to be used Tuesday; and the steel is sometimes so hot even then that it can hardly be touched. This method leaves the steel very soft and easy to work.

Steel is sometimes annealed by bringing it up to a proper heat in a furnace and then allowing the steel and the furnace to cool off together.

Annealing is done in pits by building up a fire-brick pit, filling it with steel, either in piles or packed in boxes, leaving spaces for the burning gas to circulate between the piles or boxes, covering the whole over with a fire-brick-lined cover, and heating the pit up to the proper temperature by burning gas in it. This gas is admitted through openings in the side of the pit left for the purpose. When the steel has been heated evenly to the proper temperature, the gas is turned off and the pit and its contents slowly cooled. This is the method used for annealing steel from which tin-plate is made.

The underlying principle is the same in any case —the steel is first heated uniformly to a "hardening heat" and then cooled slowly, the slower the better. Sometimes, to prevent oxidation, precautions are taken to keep the air away from the steel both during heating and cooling.

CHAPTER XI.

TOOL FORGING AND TEMPERING.

It is assumed that the general method of tempering as described before is understood, and only special directions will be given in particular cases in the following pages.

Forging Heat.—Any tool-steel forging on which there is any great amount of work to be done should have the heavy forging and shaping done at a yellow heat. At this heat the metal works easily and properly, and the heavy pounding refines the grain and leaves the steel in proper condition to receive a cutting edge. When a tool is merely to be smoothed off or finished, or forged to a very slight extent, the work should be done at a much lower heat, just above the hardening temperature.

Very little hammering should be done at any heat below the hardening temperature.

Cold-chisels.—The ordinary cold-chisel is so simple in shape that no detail directions are necessary for forging. The stock should be heated to a yellow heat and forged into shape and finished as smooth as possible. If properly forged the end, or edge, will bulge out, like Fig. 234. This should be *nicked* across with a sharp hot-chisel (but not cut off), as shown at *C*, and broken off after the tool has been hardened. This broken edge will then

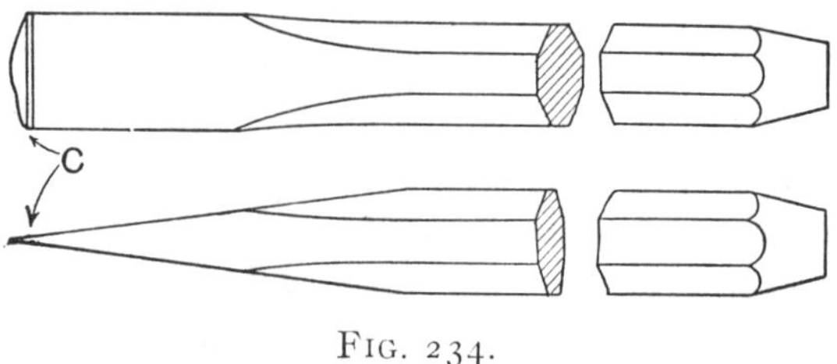

FIG. 234.

show the grain and indicate whether the steel has been hardened at a proper temperature.

When hardening, the chisel should be heated red-hot as far back from the point as the line *A*, Fig. 235. Great care must be taken to heat slowly enough to heat the thicker part of the chisel without overheating the point. If the point does become too hot, it should not be dipped in water to cool off, but allowed to cool in the air to below the hardening heat and then reheated more carefully.

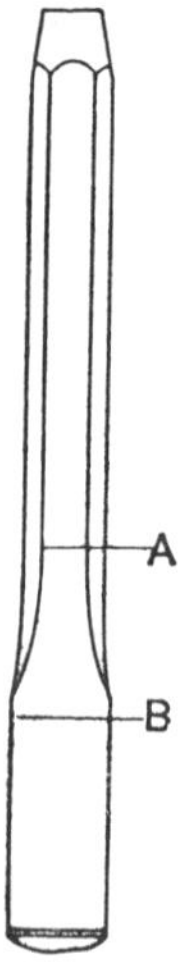

FIG. 235.

When the chisel has been properly heated to the hardening heat, the end should be hardened in cold water back to the line *B*, Fig. 235. As soon as the end is cold the chisel should be withdrawn from the water and one side of the end polished off with a piece of emery or something of that nature, as described before.

The part of the chisel from *A* to *B* will be still red-hot, and the heat from this part will gradually reheat the point of the tool. As the metal is reheated the polished surface will change color, showing at first yellow, brown, and at last purple. As soon as the purple, almost blue color reaches the nick at the end, the chisel should again be cooled, this time completely. The waste end may now be snapped off and the grain examined. To test for proper hardness, try the end of the chisel with a fine file, which should scratch it slightly. If the grain is too coarse, the tool should be rehardened at a *lower* temperature, while if the metal is too soft, it should be rehardened at a *higher* temperature.

Cape-chisel.—The cape-chisel, illustrated in Fig. 236, is used for cutting grooves and working at the

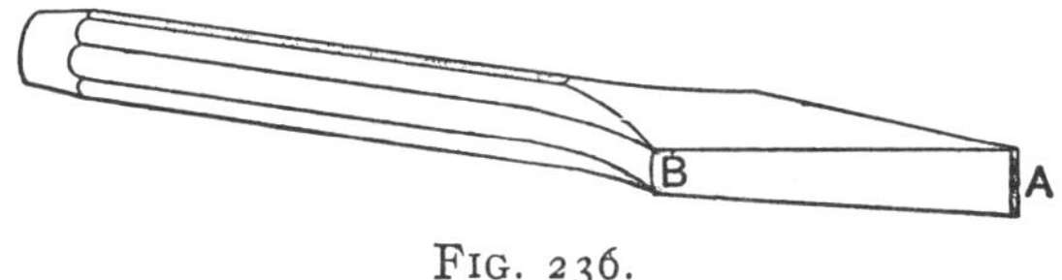

FIG. 236.

bottom of narrow channels. The cutting edge *A* should be wider than any part of the blade back to *B*, which should be somewhat thinner in order that the blade may "clear" when working in a slot the width of *A*.

The chisel is started by thinning down *B* over the horn of the anvil, as shown at *A*, Fig. 237. The finishing is done with a hammer or flatter in the manner illustrated at *B*. The chisel should *not* be worked flat on top of the anvil, as shown at *C*, as this knocks the blade out of shape.

The cape-chisel is tempered the same as a cold-chisel.

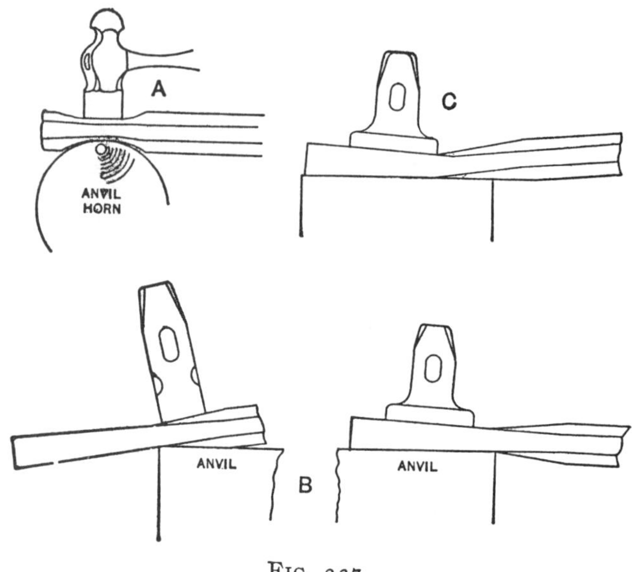

FIG. 237.

Square- and Round-nose Cape-chisels.—The chisels are started in the same way as an ordinary cape-chisel, the ends being left somewhat more stubby.

The end is then finished round or square, as shown in Fig. 238, and tempered the same as a cold-chisel.

Round-nose cape-chisels are sometimes used to center drills, and are then called "centering" chisels.

Lathe-tools in General.—The same general forms of lathe-tools are followed in nearly all shops; but in different places the shapes are altered somewhat to suit individual tastes.

Right- and Left-hand Tools.—Such tools as side tools, diamond-points, etc., are generally made in pairs—that is, right- and left-handed. If a tool is made with the cutting edge on the left-hand side (as the tool is looked at from the top with the shank

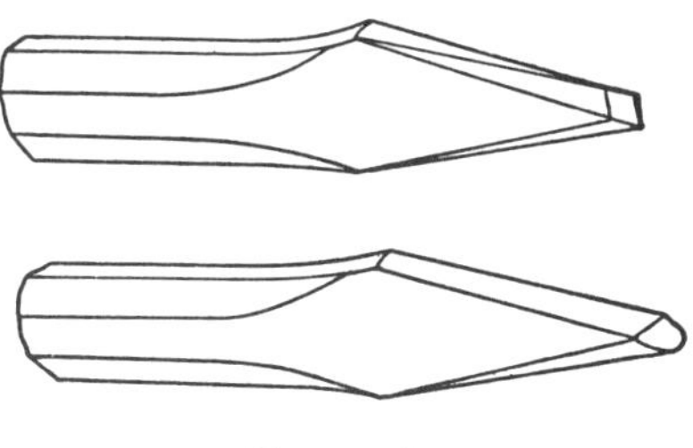

FIG. 238.

of the tool nearest the observer), it would be called a *right-hand* tool. That is, a tool which begins its cut at the right-hand end of the piece and cuts towards the left is known as a *right-hand* tool; one commencing at the left-hand end and cutting towards the right would be known as a *left-hand* tool.

The general shape of right- and left-hand tools for the same use is practically the same excepting that the cutting edges are on opposite sides.

Round-nose and Thread Tools.—Round-nose and thread tools are practically alike, the difference being in the grinding of the end. The thread tool is sometimes made a little thinner at the point.

The round-nose tool, Fig. 239, is so simple in shape that no description of the forging is necessary. Care must be taken to have proper "clearance." The cutting is all done at or near the end, and the sides must be so shaped that they "clear" the upper edge of the end. In other words, the upper edge of the shaped end must be

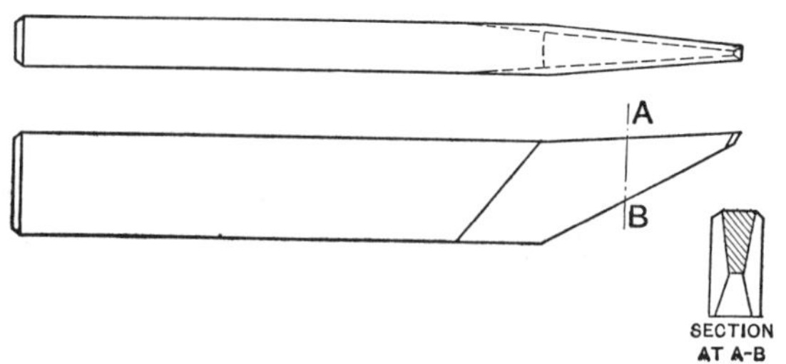

FIG. 239.

wider at every point than the lower edge, as shown by the section.

For hardening, the tool should be heated about as far as the line *A*, Fig. 240, and cooled up to the

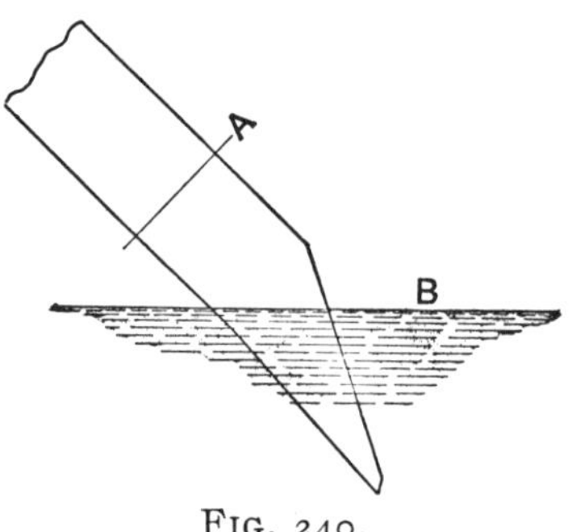

FIG. 240.

line *B*. Temper color of scale should be light yellow.

Cutting-off Tools.—Cutting-off tools are forged with the blade either on one side or in the center of the stock. The easier way to make them is to forge the blade with one side flush with the side of the tool. A tool forged this way is shown in Fig. 241.

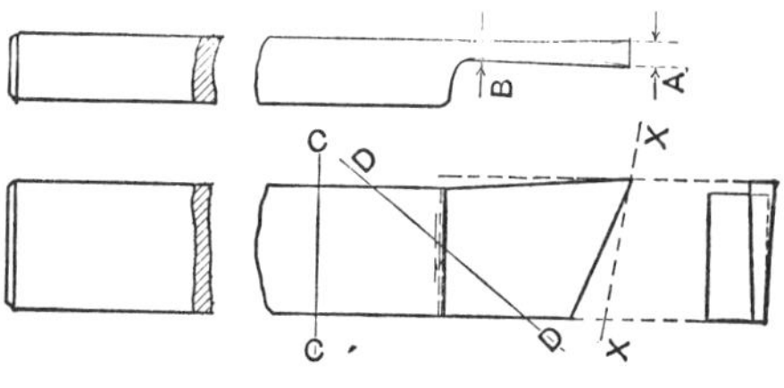

FIG. 241.

The cutting edge is the extreme tip of the blade, and the cutting is done by forcing the tool straight into the work, the edge cutting a narrow groove. The only part of the tool which should touch the work is the extreme end, or cutting edge; therefore the th ckest part of the blade must be the cutting edge, the sides gradually tapering back in all directions and becoming thinner, as shown in the drawing, *A* being wider than *B*.

The cutting edge should be slightly above the level of the top of the tool, or, in other words, the blade should slant slightly upwards.

The clearance angle at the end of the tool is about right for lathe-tools; but for plainer tools the end should be made more nearly square, about as shown by the line *X*—*X*.

For hardening, the heat should extend to about the line *C*—*C*, and the end should be cooled to about the line *D*—*D*. Temper color should be light yellow.

The tool may be forged by starting with a fuller cut, as shown at *A*, Fig. 242. The blade is roughly forged into shape with a sledge, or, on light stock, a hand-hammer, working over the edge of the anvil to form the shoulder in the manner shown at *B*. This leaves the end bulged out and in rough shape,

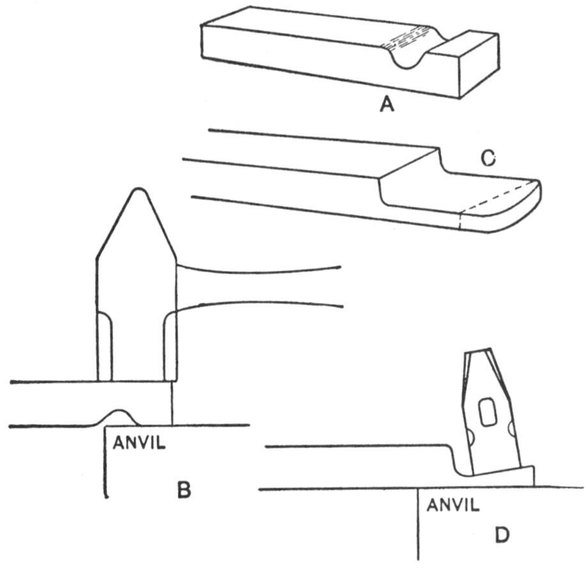

FIG 242.

similar to *C*. The end should be trimmed off with a sharp hot-chisel along the dotted line.

The finishing may be done over the corner of the anvil, using a hand-hammer or flatter, in the same way as when starting the tool; or a set-hammer may be used, as shown at *D*.

Care must be taken to have proper clearance on all sides of the blade. It is a good plan to upset the end of the blade slightly by striking a few light blows the last thing on the end at the cutting edge, then adding a little clearance.

When a tool is wanted with the blade forged in the center of the shank, the two shoulders are formed by using a set-hammer and working at the edge of the anvil face, letting the corner of the anvil shape one shoulder while the set-hammer is forming the other. This process has been described before under the general method of forming double shoulders.

Bent Cutting-off Tool.—The bent cutting-off tool,

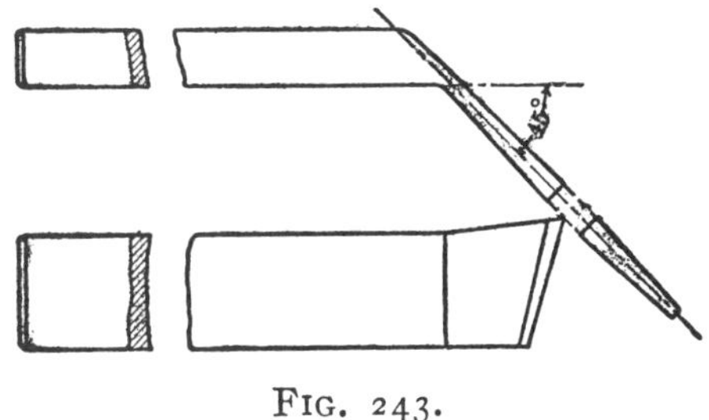

FIG. 243.

Fig. 243, is made and tempered exactly the same as the straight tool, excepting that the blade is bent backward toward the shank through an angle of about 45 degrees.

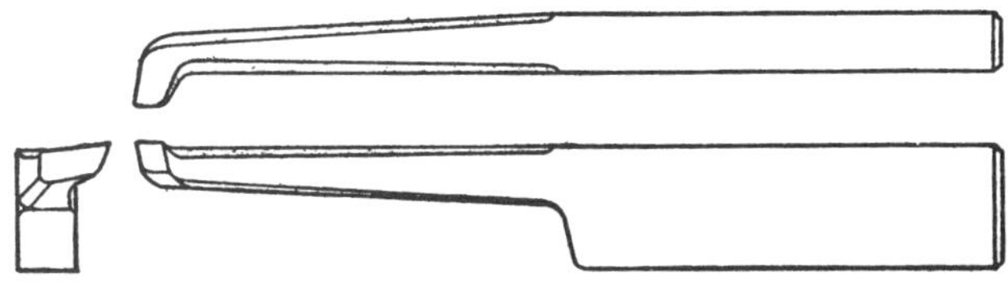

FIG. 244.

Boring Tool.—The boring tool, illustrated in Fig. 244, needs no particular description. The length of the thin shank depends upon the depth of the hole the tool is to be used in, but, as a general rule, should not be made any longer than necessary.

This thin shank is started with a fuller cut and drawn out in much the same way as the cutting-off tool was started.

The cutting edge is at the end of the small, bent nose. The only part of the tool required tempered is the bent nose, or end, which should be given the same temper color as the other lathe-tools—light yellow.

Internal Thread Tool.—This tool, used for cutting screw threads on the inside of a hole, is forged to the same shape as the boring tool described above, the end being afterward ground somewhat differently.

Diamond-points.—These tools are made right and left.

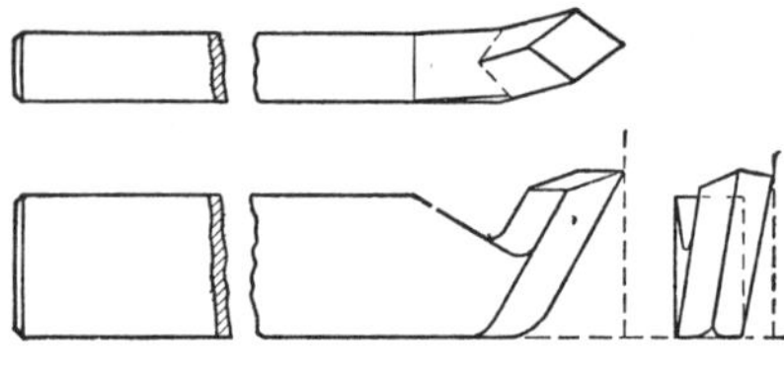

FIG 245.

There are several good methods of making these tools; but the one given below is about as quick and easy as any, and requires the use of no tools excepting the hand-hammer and sledge.

The diamond-point is started as shown at *A*, Fig. 246, by holding the stock at an angle of about 45 degrees over the outside edge of the anvil. It is first slightly nicked by being driven down with a sledge against the corner, and the bent end down to the dotted position with a few blows, as indicated by the arrow.

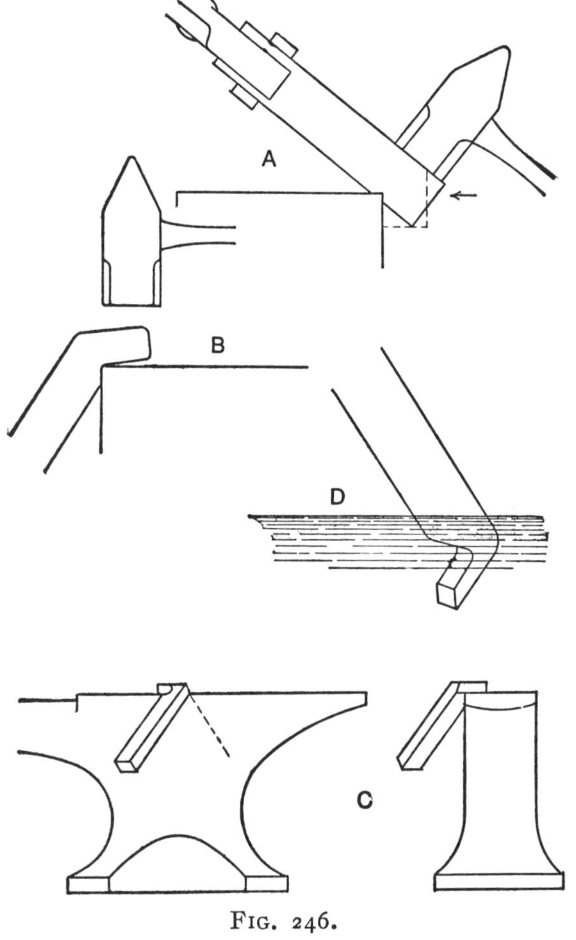

Fig. 246.

This end is further bent by holding and striking as illustrated at *B*. The diamond shape is given to the end by swinging the tool back and forth and striking as shown at *C*, which gives a side and end view of tool in position on the anvil.

The tool is finished by trimming the end with a sharp hot-chisel and so bending the end as to throw the top of the nose slightly to one side, giving the necessary side "rake" as shown in Fig. 245.

When hardening, the end should be dipped as shown at *D* and the temper drawn to show light-yellow scale.

Tools like the above made of stock as large as $\frac{1}{2}'' \times 1''$ may be made using the hand-hammer alone.

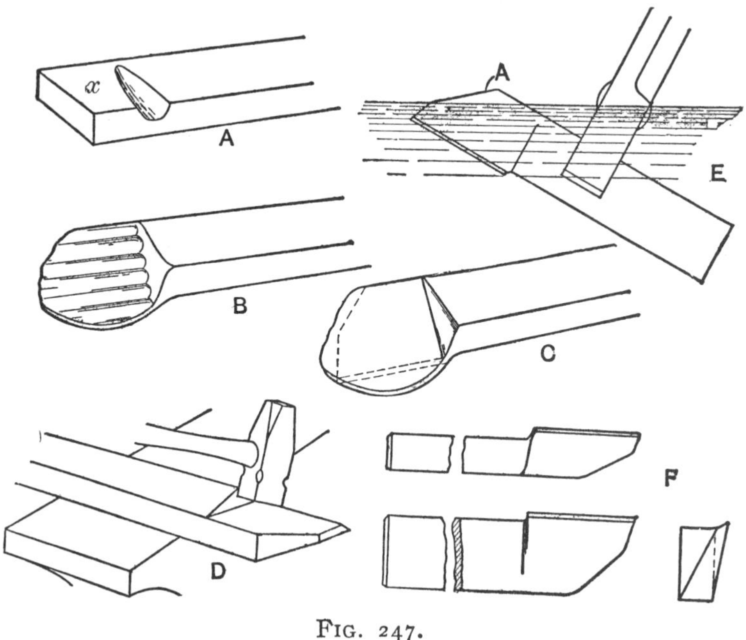

Fig. 247.

Side Tools.—Side tools, or side-finishing tools, as they are also called, are generally made about the shape shown in Fig. 247. These tools are made right and left and are also made bent. The bent side tools leave the ends forged the same; but the blade is afterward bent toward the shank, cutting edge out, at an angle of about 45 degrees.

The side tool may be started by making a fuller cut as shown at *A*, Fig. 247, near the end of the stock.

The part of the stock marked *x* is then drawn out by using a fuller turned in the opposite direction, working the stock down into the shape shown at *B*. The blade is smoothed up with a set-hammer and trimmed with a hot-chisel along the dotted lines on *C*. The curved end of the blade is smoothed up and finished with a few blows of the hand-hammer.

The tool is finished by giving the proper "offset" to the top edge of the blade. This is done by placing the tool, flat-side down, with the blade extending over, and the end of the blade next the shank about one-eighth of an inch beyond, the outside edge of the anvil. A set-hammer is placed on the blade close up to the shoulder and slightly tipped, so that the face of the hammer touches the thin edge of the blade only, as illustrated at *D*. One or two light blows with the sledge will give the necessary offset, and after straightening the blade the tool is ready for tempering.

It is very important on these tools, as well as on all others, to have the cutting edge as smooth and true as possible; it is, therefore, best, the very last thing, to smooth up this part of a tool, using the hand- or set-hammer. Above all things, the cutting edge must not be rounded off, as this necessitates grinding down the edge until the rounded part has been completely ground off.

While the side tool is being heated for tempering, it should be placed in the fire with the cutting edge up. It is more easy to avoid overheating of the edge in this way.

The blade is hardened by dipping in water as shown at *E*, only a small part of back, *A*, of the blade extending above the water and remaining red-hot. The tool is taken from the water, quickly polished on the flat side, and the temper drawn to show a very light yellow. The same color should show the entire length of the cutting edge. If the color shows darker at one end, it indicates that that end of the blade was not cooled enough, and the tool should be rehardened, this time tipping the tool in such a way as to bring that end of the blade which was before too soft deeper in the water.

Centering Tool.—The centering tool, Fig. 248, used for starting holes on face-plate and chuck work, is started in much the same way as the boring tool. The end is flattened out thin and trimmed into shape with a hot-chisel. The right-hand side of the end should be cut from the top side and the left-hand from the other, leaving the end the same shape as a flat drill.

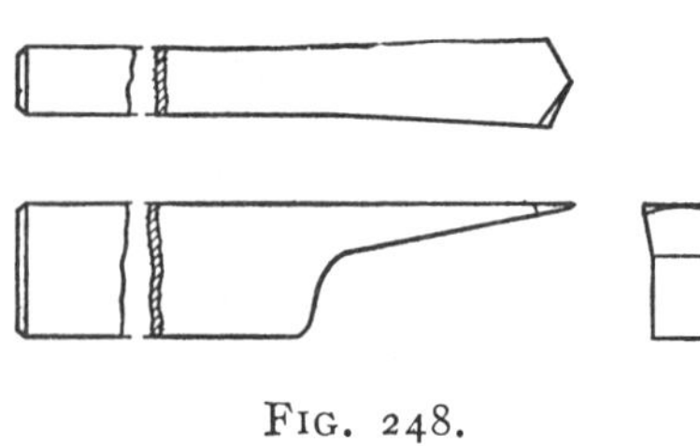

FIG. 248.

Tempered the same as other lathe-tools.

Finishing Tool.—This tool, Fig. 249, is started by

bending the end of the stock down over the edge of the anvil in the same way as when starting the diamond-point.

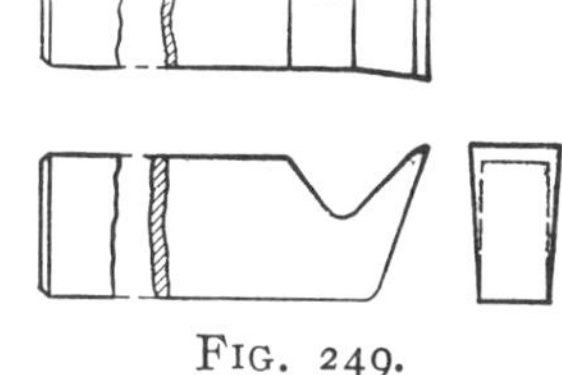
FIG. 249.

The end is flattened and widened by working with a hand- or set-hammer, as shown at *A*, Fig. 250. This leaves the end bent out too nearly straight; but, after being shaped, it is bent into the proper angle, in the manner illus-

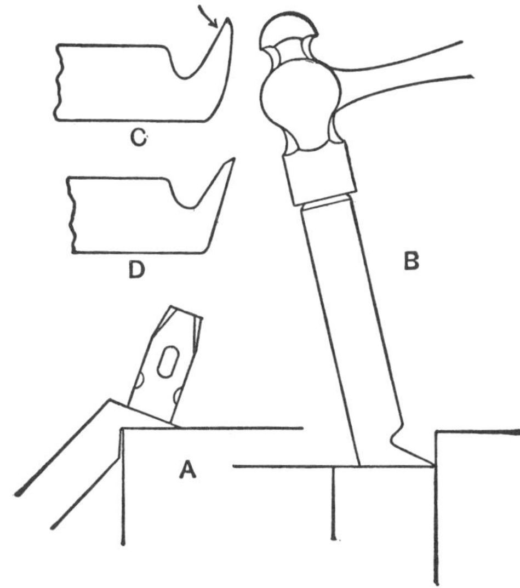

FIG. 250.

trated at *B*. The blade will then probably be bent somewhat like *C*, but a few blows with a hammer, at the point and in the direction indicated by the arrow, will straighten this out, leaving it like *D*. After trimming and smoothing, the tool is ready for tempering. The blade should be tempered to just show the very lightest yellow at cutting edge.

When a tool of this kind is to be used on a planer, the front end should make more nearly a right angle with the bottom; or, in other words, there should be less front "rake" or "clearance."

Flat Drills.—The flat drill, Fig. 251, needs no

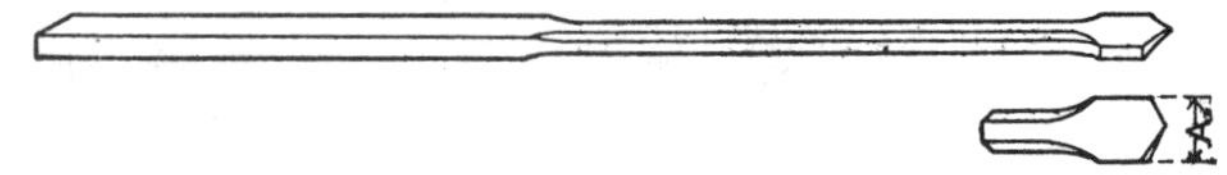

FIG. 251

description, as the forging and shaping are very simple. The end should be trimmed the same as the centering tool. The size of the tool is determined by the dimension *A*, this being the same size as the hole the drill is intended to make; thus, if this dimension were 1", the drill would be known as an inch drill.

The temper is drawn to show a dark-yellow scale.

Hammers.—As a general rule, when making hammers of all kinds by hand the eye is made first. A bar of steel of the proper size and convenient length for handling is used, and the hammer forged on the end, as much forging and shaping as possible being done before cutting the hammer from the bar.

The hole for the eye is punched in the usual way at the proper distance from the end of the bar, using a punch having a handle (Fig. 70).

The nose of the punch is slightly smaller but has the same shape as the eye is to finish. Great care must be taken to have the hole true and straight. It is very difficult and sometimes impossible to straighten up a crooked hole.

After punching the eye, the sides of the stock are generally bulged out, and to prevent knocking the eye out of shape while forging down this bulge a drift-pin, Fig. 252, is used. This is made

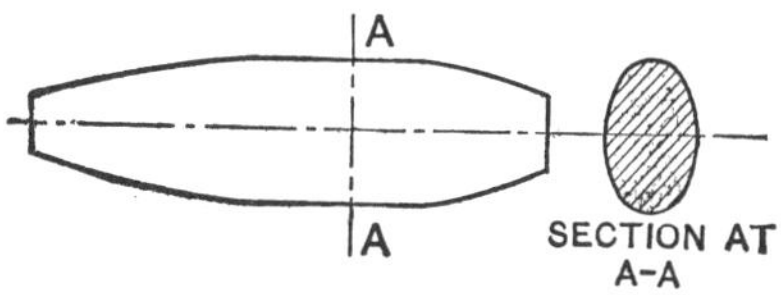

FIG. 252.

of tool-steel and tapers from near the center toward each end, one end being somewhat smaller than the other. The center of the pin is the same shape and size as the eye is to be in the hammer.

When the bar has been heated the drift-pin is driven tightly into the hole and the bulge forged down in the same way (*B*, Fig. 254) as a solid bar would be treated. When the drift-pin becomes heated it must be driven out and cooled, and under no circumstances should the bar be heated with the pin in the hole. The pin should always be used when there is danger of knocking the eye out of shape.

The steel used for hammers, and "battering tools" in general, should be of a lower temper (contain less carbon) than that used for lathe-tools.

The eye of a hammer should not be of uniform size throughout, but should be larger at the ends and taper slightly toward the center, as illustrated in Fig. 253, which shows a section of a hammer cut through the center of the eye. When the eye is made in this way (slightly contracted at the middle), the hammer-handle may be driven in tightly from one end; then by driving one or more wedges in the end of the handle it is held firmly in place and there is no chance for the head to work up or down.

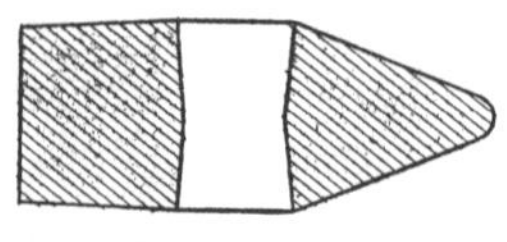

FIG. 253.

Cross-pene, Blacksmith's or Riveting Hammer.—A hammer of this kind is shown at *C*, Fig. 6.

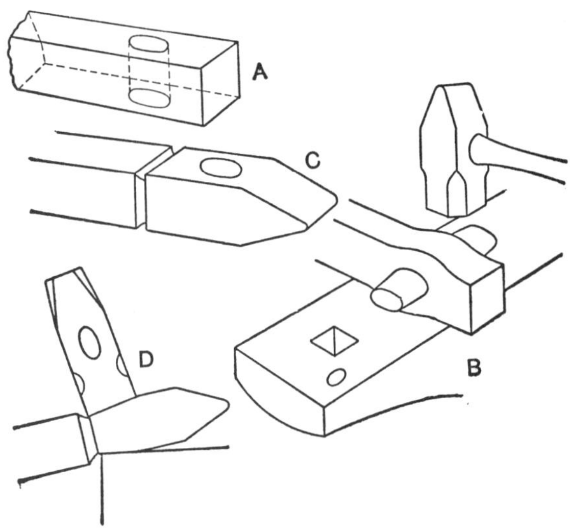

FIG. 254.

The different steps in the process of forging are illustrated in Fig. 254. First the eye is punched as shown at *A*. The pene is then drawn out and shaped and a cut started at the point where the end of the hammer will come (*C*), the drift-pin being used, as shown at *B*, while forging the metal around the eye.

The other end of the hammer is then worked up into shape, using a set-hammer as indicated at *D*.

When the hammer is as nearly finished as may be while still on the bar, it is cut off with a hot-chisel, leaving the end as nearly square and true as possible.

After squaring up and truing the face the hammer is tempered.

For tempering, the whole hammer is heated in a slow fire to an even hardening heat; while hardening, the tongs should grasp the side of the hammer, one jaw being inserted in the eye.

Both ends should be tempered, this being done by hardening first one end, then the other.

The small end is hardened first by cooling, as shown in Fig. 255. As soon as this end has cooled, the position is instantly reversed and the large end of the hammer dipped in the water and hardened. While the large end is cooling, the smaller one is polished and the temper color watched for. When a dark-brown scale appears at the end the hammer is again reversed, bringing the large end uppermost and the pene in the water. The face

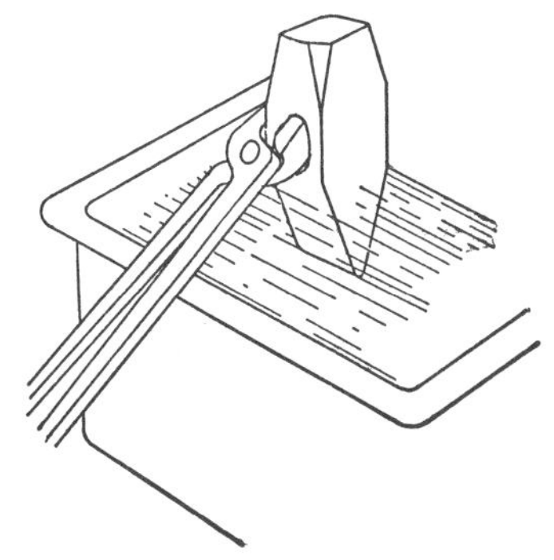

FIG. 255.

end is polished and tempered in the same way as the small end. If the large end is properly hardened before the temper color appears on the small end, the hammer may be taken completely out of the water and the large end also polished, the colors being watched for on both ends at once. As soon as one end shows the proper color it is promptly dipped in water, the other end following as soon as the color appears there.

Under no circumstances should the eye be cooled while still red-hot.

For some special work hammer-faces must be very hard; but for ordinary usage the temper as given above is very satisfactory.

Ball Pene-hammer.—The ball pene-hammer, Fig. 5, is started by punching the eye.

The hammer is roughed out with two fullers in the manner illustrated at *A*, Fig. 256.

The size of stock used should be such that it will easily round up to form the large end of the hammer.

After the hammer is roughed out as shown at *A*, the metal around the eye is spread sidewise, using two fullers as illustrated at *B*, a set-hammer being used for finishing. This leaves the forging like *C*. The next step is to round and shape the ball, which is forged as nearly as possible to the finished shape.

After doing this a cut is made in the bar where the face of the hammer will come, and the large end rounded up, leaving the work like *D*.

The necked parts of the hammer each side of the eye are smoothed and finished with fullers of the proper size. Some hammers are made with these necks round in section, but the commoner shape is octagonal.

After smoothing off, the hammer is cut from the bar and the face forged true. Both ends are ground true and tempered. This hammer should be tempered in the same way as described above for tempering the riveting-hammer.

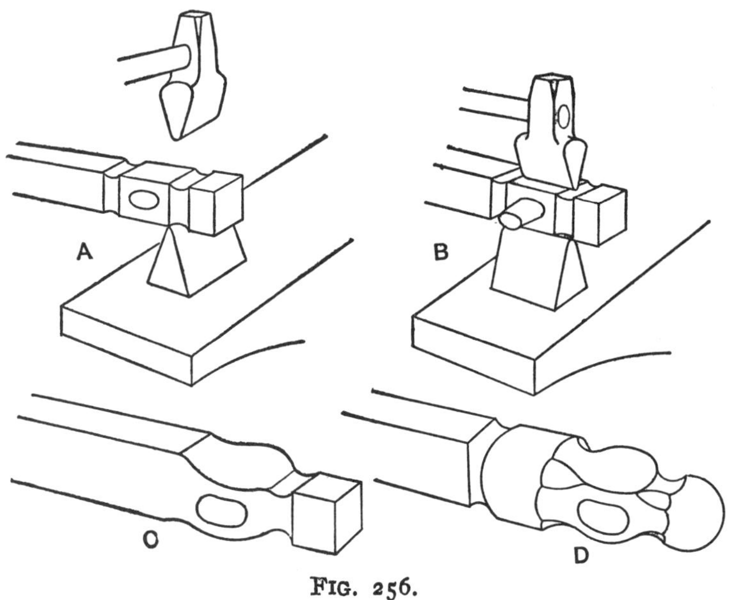

FIG. 256.

Ball pene-hammers may be made with the steam-hammer in practically the same way as described, only substituting round bars of steel for use in place of fullers.

Sledges.—Sledges are made and tempered in the same general way as riveting-hammers. Sledges may be forged and finished almost entirely under the steam-hammer.

Blacksmith's Cold-chisel.—This tool (Fig. 2) is forged in practically the same way as the cross-pene hammer described before. The end, of course, is drawn out longer and thinner, the thin edge coming parallel with the eye instead of at right angles to it.

The *cutting edge* only of the chisel is tempered. The temper should be drawn to show a bluish scale just tinged with a little purple. Under no circumstances should the head of the chisel be hardened, as this would cause the end to chip when in use and might cause a serious accident.

The tool shown in Fig. 192 may be used to advantage when making hot- or cold-chisels with the steam-hammer. By using this tool, as illustrated in Fig. 193, the blade of the chisel may be quickly drawn out and finished.

Hot-chisel.—After forming the eye of the hot-chisel (Fig. 2), the blade is started by making the two fuller cuts, as illustrated in Fig. 257. The end is drawn down as indicated by the dotted lines. The head is shaped and the chisel cut from the bar in the same way that the riveting-hammer was treated.

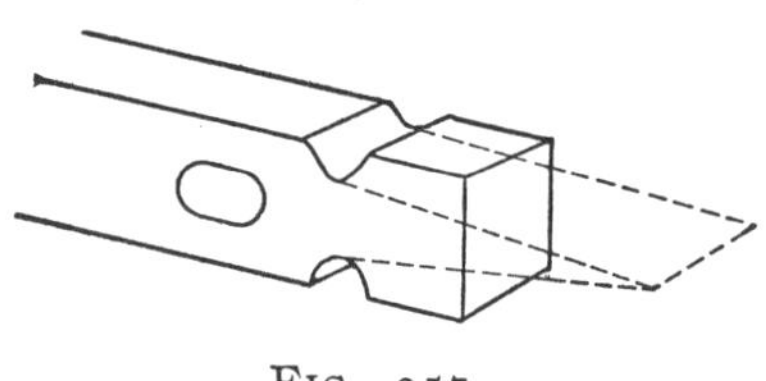

FIG. 257.

This chisel should have its cutting edge tempered the same as that of the cold-chisel.

Hardies.—Hardies such as shown in Fig. 2 should be started by drawing out the stem. This stem is drawn down to the right size to fit the hardy-hole in the anvil and the piece cut from the bar. This is heated, the stem placed in the hole in the anvil, and the piece driven down into the hole and against the face of the anvil, thus forming a good shoulder between the stem and the head of the hardy.

After forming the shoulder, the blade is worked out, starting by using two fullers in the same way as when starting the hot-chisel blade.

The cutting edge should be given the same temper as a cold-chisel.

Blacksmith's Punches. — Punches shaped similar to Fig. 70 are started the same manner as the hot-chisel, excepting that the fuller cuts are made on four sides, as shown in Fig. 258. The end is then drawn out to the shape shown by the dotted lines.

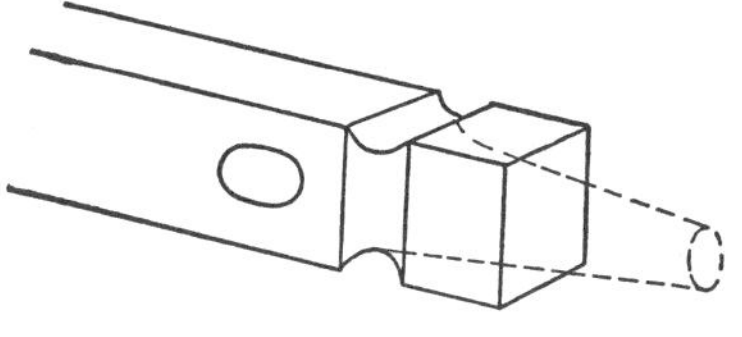

FIG. 258.

Temper same as cold-chisel.

Set-hammers — Flatters. — The set-hammer, Fig. 15, is so simple that no directions are necessary for shaping. The face only should be tempered and that should show a dark-brown or purple color.

Flatters such as shown in Fig. 14 may be made by upsetting the end of a small bar, the upset part forming the wide face; or a bar large enough to form the face may be used and the head, or shank, drawn down.

The eye should be punched after the face has been

made. The face should be tempered to about a blue.

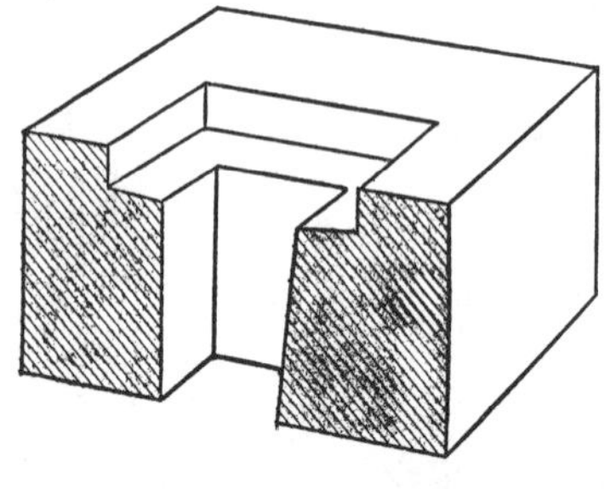

FIG. 259.

When many are to be made a swage-block similar to Fig. 259 should be used. Half only of the block is shown in the figure, the other half being cut away to show the shape of the hole which is the size of the finished flatter.

When using this block the stock is first cut to the proper length, heated, placed in the hole, and upset.

Swages.—Swages may also be made in a block similar to the one used for the flatter. The swage should be first upset in the block and the crease formed the last thing. The crease may be made with a fuller or a bar of round stock the proper size.

Fullers.—Fullers are made in the same way as swages.

All of these tools may be upset and forged under the steam-hammer, using the die, or swage, blocks as described above. The swage-blocks may be made of cast iron.

CHAPTER XII.

MISCELLANEOUS WORK.

Shrinking.—When iron is heated it expands, and upon being cooled it contracts to practically its original size.

This property is utilized in doing what is known as "shrinking."

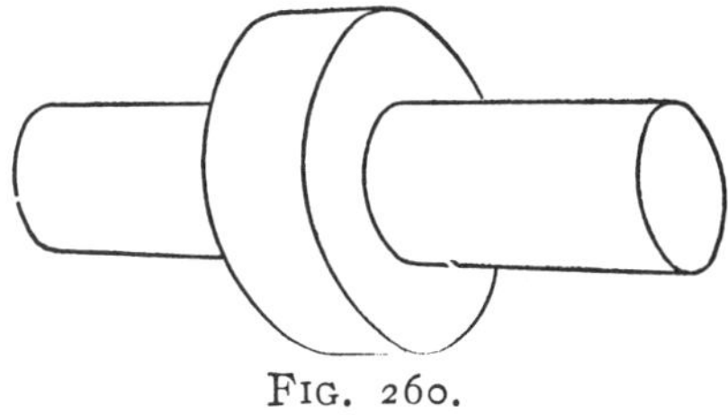

FIG. 260.

A common example of this sort of work is illustrated in Fig. 260, showing a collar "shrunk" on a shaft. The collar and shaft are made separately. The inside diameter of the hole through the collar is made slightly less than the outside diameter of the shaft. When the collar and shaft are ready to go together the collar is heated red-hot. The high temperature causes the metal to expand and thus increases the diameter of the hole, making it larger (if the sizes have been properly proportioned) than the outside diameter of the shaft. The collar is then taken from the fire, brushed clean of all ashes and dirt, and slipped on the shaft and into the proper position, where it is cooled as quickly as possible. This cooling causes the collar to contract and locks it firmly in place.

If the collar be allowed to cool slowly it will heat the shaft, which will expand and stretch the collar somewhat; then, as both cool together and contract, the collar will be loose on the shaft.

This is the method used for shrinking tires on wheels. The tire is made just large enough to slip on the wheel when hot, but not large enough to go on cold. It is then heated, put in place, and quickly cooled.

Couplings are frequently shrunk on shafts in this way.

Brazing.—Brazing, it might be said, is soldering with brass.

Briefly the process is as follows: The surfaces to be joined are cleaned thoroughly where they are to come in contact with each other. The pieces are then fastened together in the proper shape by binding with wire, or holding with some sort of clamp. The joint is heated, a flux (generally borax) being added to prevent oxidation of the surfaces, and the "spelter" (prepared brass) sprinkled over the joint, the heat being raised until the brass melts and flows into the joint, making a union between the pieces. Ordinarily it requires a bright-red or dull-yellow heat to melt the brass properly.

Almost any metal that will stand the heat can be brazed. Great care must be used when brazing cast iron to have the surfaces in contact properly cleaned to start with, and then properly protected from the oxidizing influences of the air and fire while being heated.

Brass wire, brass filings, or small strips of rolled brass may be used in place of the spelter. Brass wire in particular is very convenient to use in some places, as it can be bent into shape and held in place easily.

A simple brazed joint is illustrated in Fig. 261, which shows a flange (in this case a large washer) brazed around the end of a pipe. It is not necessary to use any clamps or wires to hold the work together, as the joint may be made tight enough to hold the pieces in place. The joint should be tight enough in spots to hold the pieces together, but must be open enough to allow the melted brass to run between the two pieces. Where the pipe comes in contact with the flange the outside should be free from scale and filed bright, the inside of the flange being treated in the same way.

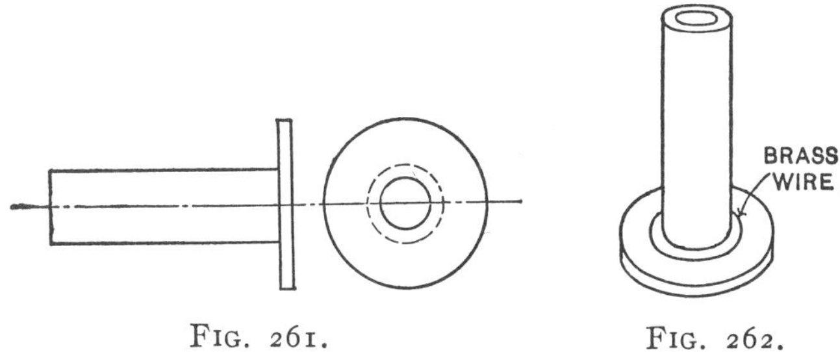

FIG. 261. FIG. 262.

When the pieces have been properly cleaned and forced together, a piece of brass wire should be bent around the pipe at the joint, as shown in Fig. 262, and the work laid on the fire with the flange down. The fire should be a clean bright bed of coals. As soon as the work is in the fire the joint should be sprinkled with the flux; in fact, it is a good plan to put on some of the flux *before* putting the work in the fire. Ordinary borax can be used as a flux, although a mixture of about three parts borax and one part sal ammoniac seems to give much better results.

The heat should be gradually raised until the brass melts and runs all around and into the joint, when the piece should be lifted from the fire.

The brazing could be done with spelter in place of the brass wire. If spelter were to be used the piece would be laid on the fire and the joint covered with the flux as before. As soon as the flux starts to melt, the spelter mixed with a large amount of flux is spread on the joint and melted down as the brass wire was before. For placing the spelter when brazing it is convenient to have a sort of a long-handled spoon. This is easily made by taking a strip of iron about $\frac{3}{4}'' \times \frac{1}{8}''$ three or four feet long and hollowing one end slightly with the pene end of the hammer.

There are several grades of spelter which melt at different heats. Soft spelter melts at a lower heat than hard spelter, but does not make as strong a joint.

Spelter is simply brass prepared for brazing in small flakes and can be bought ready for use. The following way has been recommended for the preparation of spelter: Soft brass is melted in a ladle and poured into a bucket filled with water having in it finely chopped straw, the water being given a swirling motion before pouring in the brass. The brass settles to the bottom in small particles. Care must be taken when melting the brass not to burn out the zinc. To avoid this, cover the metal in ladle with powdered charcoal or coal. When the zinc begins to burn it gives a brilliant flame and dense white smoke, leaving a deposit of white oxide of zinc.

Another example of brazing is the T shown in

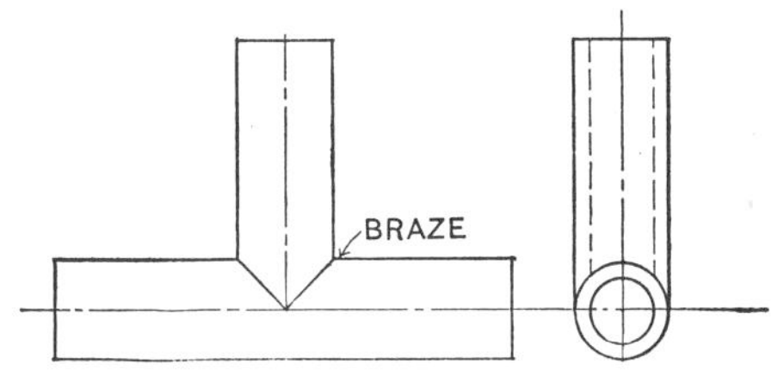

FIG. 263.

Fig. 263. Here two pipes are to be brazed to each other in the form of an inverted T.

A clamp must be used to hold the pieces in proper position while brazing, as one pipe is simply stuck on the outside of the other. A simple clamp is shown in Fig. 264 consisting of a piece of flat iron having one hole near each end to receive the two small bolts, as illustrated. This strip lies across the end of the pipe forming the short stem of the T, and the bent ends of the bolts hook

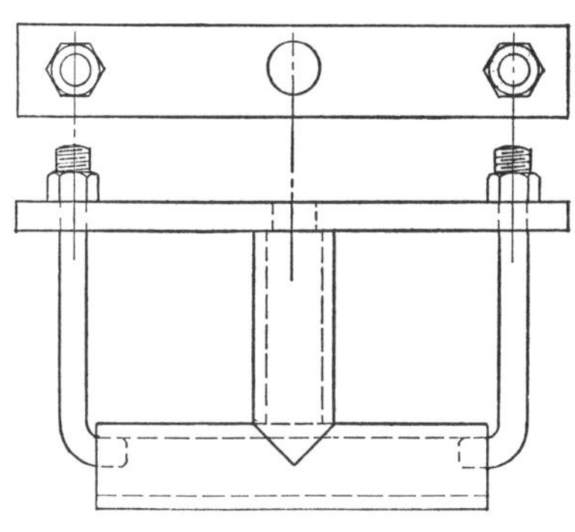

FIG. 264.

into the ends of the bottom pipe. The whole is held together by tightening down on the nuts.

The brazing needs no particular description, as the spelter or wire is laid on the joint and melted into place as before.

A piece of this kind serves as a good illustration of the strength of a brazed joint. If a well-made joint of this kind be hammered apart, the short limb will sometimes tear out or pull off a section of the longer pipe, showing the braze to be almost as strong as the pipe.

When using borax as a flux the melted scale should be cleared (or scraped) from the work while still red-hot, as the borax when cold makes a hard, glassy scale which can hardly be touched with a file. The cleaning may be easily done by plunging the brazed piece, while still red-hot, into water. On small work the cleaning is very thoroughly done if the piece, while still red-hot, is dipped into melted cyanide of potassium and then instantly plunged into water. If allowed to remain in the cyanide many seconds the brass will be eaten off and the brazing destroyed.

It is not always necessary when brazing wrought iron or steel to have the joint thoroughly cleaned; for careful work the parts to be brazed together should be bright and clean, but for rough work the pieces are sometimes brazed without any preparation whatever other than scraping off any loose dirt or scale.

Pipe-bending. — There is one simple fact about pipe-bending which, if always carried in mind, makes it comparatively easy.

Let the full lines in Fig. 265 represent a cross-section of a piece of pipe before bending. Now suppose the pipe be heated and an attempt made to bend it without taking any precautions what-

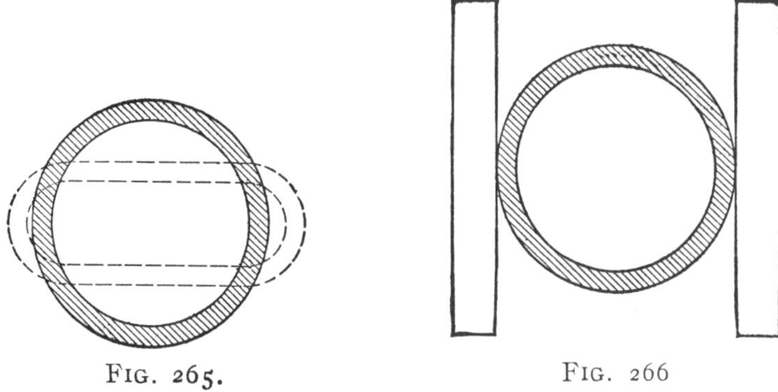

FIG. 265. FIG. 266

ever. The concave side of the pipe will flatten down against the outside of the curve, leaving the cross-section something as shown by the dotted lines; that is, the top and bottom of the pipe will be forced together, while the sides will be pushed apart. In other words, the pipe collapses.

If the sides can be prevented from bulging out while being bent it will stop the flattening together of the top and bottom. A simple way of doing this is to bend the pipe between two flat plates held the same distance apart as the outside diameter of the pipe (Fig. 266). Pipe can sometimes be bent in a vise in this way, the jaws of the vise taking the place of the flat plates mentioned above.

Large pipe may be bent something in the following way: If the pipe is long and heavy the part to be bent should be heated, and then while one end is supported, the other end is dropped repeatedly on the floor. The weight of the pipe will cause it to bend in the heated part. Fig. 267 illustrates this,

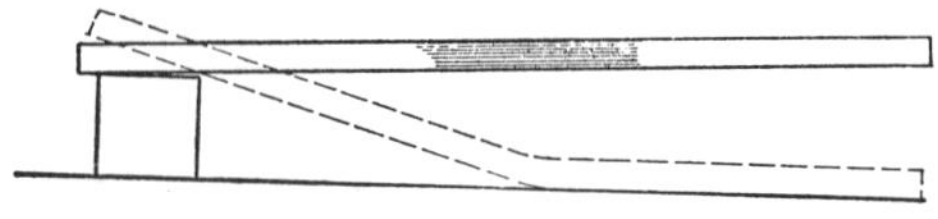

FIG. 267.

the solid lines showing the pipe as it is held before dropping and the dotted lines the shape it takes as it is dropped.

As the pipe bends the sides, of course, bulge out, and the top and bottom tend to flatten together; but this is remedied by laying the pipe flat and driving the bulging sides together with a flatter.

Another way of bending is to put the end of the pipe in one of the holes of a heavy swage-block (as illustrated in Fig. 268), the bend then being made

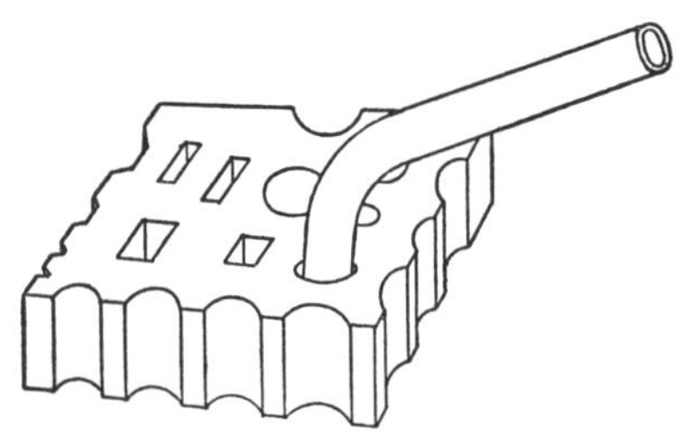

FIG. 268.

by pulling over the free end. The same precaution must, of course, be taken as when bending in other ways.

The fact that by preventing the sides of the pipe from bulging it may be made to retain its proper shape is particularly valuable when several pieces are to be bent just alike. In this case a jig is made which consists of two side plates, to prevent the sides of the pipe from bulging, and a block between these plates to give the proper shape to the curve.

A piece of bent pipe which was formed in this

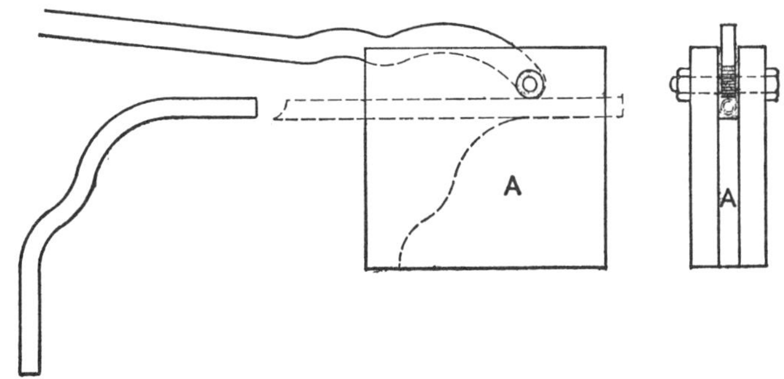

FIG. 269.

way is shown in Fig. 269, together with the jig used for bending it.

The pipe was regular one-quarter-inch gas-pipe. The jig was made as follows: The sides were made of two pieces of board about $1\frac{1}{2}''$ thick. Between these sides was a board, *A*, sawed to the shape of the inside curve of the bent pipe. This piece was slightly thicker than the outside diameter of the pipe (about $^1/_{32}''$ being added for clearance). The inside face of the sides and the edge of the block *A* were protected from the red-hot pipe by a thin sheet of iron tacked to them.

A bending lever was made by bending a piece of $\frac{3}{8}'' \times 1''$ stock into the shape of the outside of the pipe. This lever was held in place by a $\frac{1}{2}''$ bolt passing through the sides of the jig, as shown.

To bend the pipe it was heated to a yellow heat, put in the jig as indicated by the dotted lines and the lever pulled over, forcing the hot pipe to take the form of the block.

A jig of this sort is easily and cheaply made and gives good service, although it is necessary sometimes to throw a little water on the sides to prevent them from burning.

Another common way of bending is to fill the pipe with sand. One end of the pipe to be bent is plugged either with a cap or a wooden plug driven in tightly. The pipe is filled *full* of sand and the other end closed up tight. The pipe may then be heated and bent into shape. It is necessary to have the pipe *full* of sand or it will do very little good.

For very thin pipe the best thing is to fill with melted rosin. This, of course, can only be used when the tubing or pipe is very thin and is bent cold, as heating the pipe would cause the rosin to run out.

Thin copper tubing may be bent in this way.

A quite common form of pipe-bending jig is illustrated in Fig. 270.

The outside edge of the semicircular casting has a groove in it that just fits half-way round the pipe. The small wheel attached to the lever has a corresponding groove on its edge. When the two are in the position shown the hole left between them is the same shape and size as the cross-section of the pipe.

To bend the pipe, the lever is swung to the extreme left, the end of the heated pipe inserted in the catch at *A* (which has a hole in it the same size as the pipe), and the lever pulled back to the right, bending the pipe as it goes.

The stem on the lower edge of the casting was

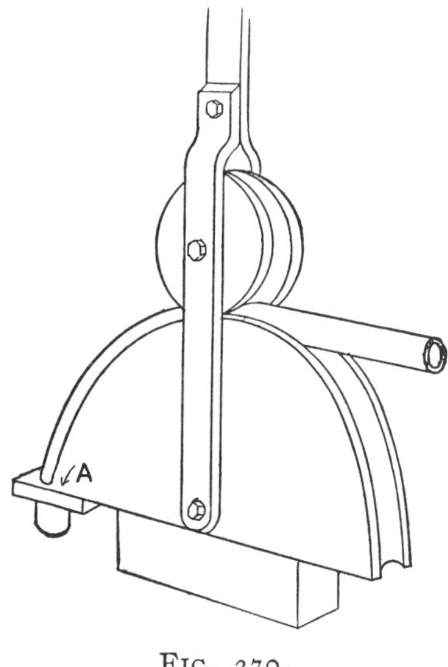

FIG. 270.

made to fit in a vise, where the jig was held while in operation.

Annealing Copper and Brass. — Brass or copper may be softened or annealed by heating the metal to a red heat and cooling suddenly in cold water, copper being annealed in the same way that steel is hardened. Copper annealed this way is left very soft, somewhat like lead. Hammering copper or brass causes it to harden and become springy. When working brass or copper, if much bending or hammering is done, the metal should be annealed frequently.

Bending Cast Iron.—It is sometimes necessary to straighten castings which have become warped or twisted. This may be done to some extent by heating the iron and bending into the desired shape. The part to be bent should be heated to what might be described as a dull-yellow heat. The bending is done by gradually applying pressure, not by blows. For light work two pairs of tongs should give about the right amount of leverage for twisting and bending.

When properly handled (very "gingerly"), thin castings may be bent to a considerable extent. Before attempting any critical work some experimenting should be done on a piece of scrap to determine at just what heat the iron will work to the best advantage, and how much bending it will stand without breaking.

Case-hardening.—Case-hardening is a process by which articles made of soft steel or wrought iron are given a hard wearing surface.

Wrought iron or machine-steel will not harden to any appreciable extent, and this fact would prevent the use of either metal in many places where they would be the ideal materials if they could only be given a hard surface.

This hard surface can, however, be had by means of the process known as "Case-hardening." (Wrought iron and machine-steel are taken as being practically alike, as they are, so far as chemical composition is concerned.)

Practically the only difference between wrought iron and tool-steel is that tool-steel contains a little more of the element called carbon than wrought iron does. Tool-steel can be hardened by heating to a red heat and cooling suddenly, because it does contain this carbon; while wrought iron can *not* be hardened, on account of the lack of it. If then by some means carbon can be added to the metal on the outside of an article made of wrought iron or machine-steel, the outside part will be made into tool-steel, and may then be hardened in the ordinary manner, while the inside metal will be soft and unchanged.

Wrought iron or machine-steel if heated to a high heat in contact with charred leather, ground bone, or other material containing a great deal of carbon will "take up" or absorb carbon from that material; and the outside will be converted into a high-carbon (tool) steel, which may be hardened by cooling suddenly from a high heat. This process is known as case-hardening, and is used for work which requires a hard wearing surface backed up by a softer and tougher material to resist shocks.

If a piece of wrought iron which has been case-hardened be broken across, it will appear something like Fig. 271. The outside layer, or coating, of hard steel can be easily distinguished from the inner core of softer unchanged metal.

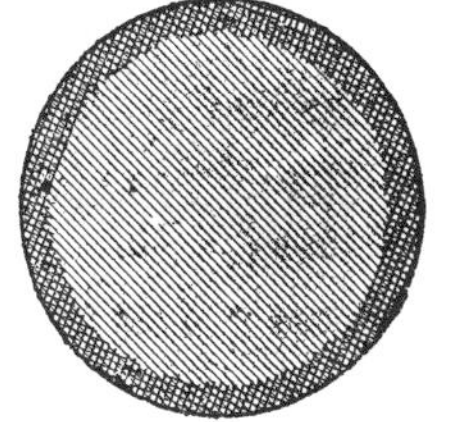

FIG. 271.

The "depth of penetration" of the carbon (or, in other words, the depth to which the iron is changed into steel) is determined by the temperature to which the metal is heated, the length of time it is kept at that temperature, and the substance it is heated in contact with.

The carbon penetrates faster at a high heat; but a high heat can not always be used, particularly if that mottled appearance so often seen on case-hardened articles is wanted. Pieces which are to be mottled should not be heated much above a good red heat, as a higher heat destroys the color. The longer the work is kept at the proper heat the deeper the carbon penetrates. So, when the same heat and the same case-hardening mixture are used all the time, the depth of hardness on the pieces treated can be determined by the length of time the pieces are hot. For ordinary work, where it is not necessary to have the mottled coloring on the pieces, about a good yellow heat should be used—about as high a heat as ordinary cast iron will stand without danger of going to pieces.

As stated before, a case-hardened piece of iron or machine-steel is really made up of two distinct metals—the outside hard shell of high-carbon steel (which is the same as tool-steel) and the inside softer core of the original material. This outside coating can be treated in the same manner as ordinary tool-steel; that is, it can be hardened and annealed just as tool-steel is. In fact, when a case-hardened article is suddenly cooled from a red heat exactly the same thing is done as when hardening a piece of ordinary tool-steel, with, however, this difference: in hardening tool-steel the piece is hardened clear through, and for ordinary purposes is so hard as to be almost useless; while with a piece of case-hardened machine-steel, for instance, the *outside only* is hardened (because the outside only is tool-steel), while the inside is left tough and comparatively soft. In the case-hardened piece there is a tough inside core which will stand shocks that would snap off a piece of hardened tool-steel and an outside coating which is as hard and is the same as hardened tool-steel. This gives a combination of hardness and toughness which is not possible with either machine-steel or tool-steel alone; and it is this fact which makes case-hardened articles so valuable for many purposes.

To repeat: The object of case-hardening is to convert the outside of a low carbon steel or iron into a high-carbon steel that can be hardened easily; and this converting is done by heating the piece in contact with some substance containing a large amount of carbon.

By taking certain precautions while heating and cooling, the surface of the case-hardened pieces may be given a mottled coloring of reds, blues, and greens, which when rightly done is sometimes very beautiful. Such coloring is often seen on gun-locks, finished wrenches, etc.

Two common methods of case-hardening are in use—what might be called the cyanide method and the bone or animal-charcoal process.

Cyanide Case-hardening.—For the cyanide method cyanide of potassium is used—the purer the better.

One way of using the cyanide is as follows: Small pieces and pieces which need only a *very thin* shell of hard steel are heated to a high red heat, drawn from the fire and sprinkled over with cyanide of potassium, reheated for a few seconds to give the carbon from the cyanide a chance to "soak in," drawn from the fire again, sprinkled with the cyanide, and cooled in cold water. This is an easy and quick way when it will answer the purpose.

This method is also of use when case-hardening in spots. When only a hole or some small part of a piece is wanted hardened a small piece of the cyanide may be placed on that particular spot and the hardening confined to the area covered by the cyanide.

Another method of using the cyanide is to melt it and heat red-hot in a ladle or pot. The pieces to be case-hardened are heated and placed in the red-hot cyanide and left there for some minutes. After "soaking" a proper length of time they are hardened by cooling in cold water.

This method when properly carried out gives a mottled appearance to the case-hardened surfaces if they have been previously polished.

The longer the articles are left in the heated cyanide the deeper will be the penetration of the carbon, although not quite in proportion to the time. Heating in this way for about ten minutes will give a penetration of perhaps one hundredth of an inch.

Case-hardening with Bone.—This method is used when a deeper coating is needed. The pieces are first packed in an iron box in such a way that they are completely surrounded by ground bone or some other material containing a great deal of animal carbon. On the bottom of the box is placed a layer of the ground bone about an inch deep; on this are laid pieces to be case-hardened, leaving a space about three-quarters of an inch wide on all sides of every piece; over these pieces is put more bone, covering them about one inch deep; then more pieces and more bone until the box is full. There must be a top layer of bone at least as thick, if not somewhat thicker, than the bottom layer. The box is then sealed up air-tight (to prevent the oxidation of the bone) with fire-clay and it and its contents heated in a furnace to the right heat and kept at this heat for several hours. The deeper the coating is needed the longer the box is kept hot. When the box has been heated long enough it is withdrawn from the fire, the top taken off, and the pieces picked out while still red-hot and hardened in cold water. Or, as is often done, after taking off the top of the box it is turned bottom side up over the tank and the whole contents, bone and all, dumped into the water.

The boxes used are made of cast or wrought iron, but cast-iron boxes are very satisfactory and are easily replaced as they wear out.

The bone may be used several times before it is "spent."

When it is desired to give the articles a bright mottled color on the surface they must be polished before case-hardening and should be packed in "charred" bone; that is, fresh bone which has been heated just hot enough to char it black.

When cooling, the cover of the box should be

removed and the articles dumped by turning the box upside down over the cooling-tank, keeping the box very close to the surface of the water.

Good results are obtained by using a mixture of about half and half charred bone and powdered charcoal.

The penetration of the carbon is perhaps one hundredth of an inch for each hour the pieces are left hot. It is possible to convert a piece of wrought iron to tool-steel clear through in this way.

Sometimes cyanide is mixed with the bone and used as above. This hastens the penetration of the carbon somewhat.

Milling-cutters and tools of that character which are only to be used once, or on light work, may be made of machine-steel and case-hardened by using bone.

After case-hardening, or carbonizing, they may be hardened, tempered, and ground in the usual way.

Case-hardening, in Parts Only, of Pieces.—Sometimes it is desirable to case-harden only certain parts of a piece. In such a case the parts to be left untreated may be covered with a coating of clay held in place with wire. Wherever the work is protected by the clay it remains uncarbonized, while the uncovered parts can be hardened. After covering the parts with the clay as above, the case-hardening may be done in the usual way by packing the object in bone and heating as usual.

Another way of obtaining the same result is to carbonize the entire surface and then machine off the parts wanted untreated; thus, suppose a shaft is wanted similar to Fig. 272, *A*, with only the parts

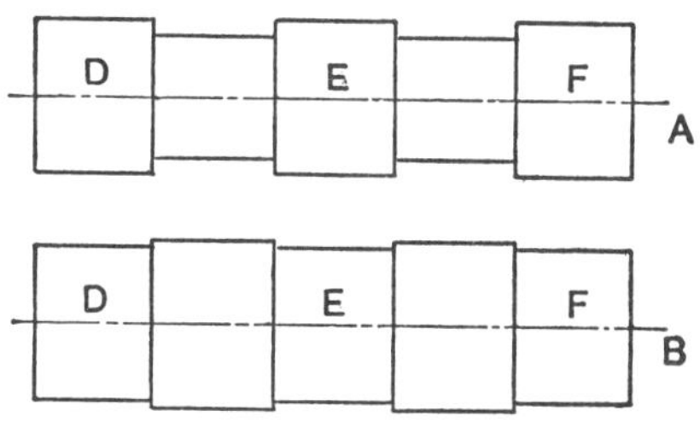

FIG. 272.

marked *D*, *E*, and *F* case-hardened. When the shaft is first made, only the parts wanted hard (*D*, *E*, and *F*) should be turned to size, the rest being left in the rough.

The shaft is then carbonized in the usual way and *annealed* instead of hardened. The rough parts are then turned to size; the cut taken removes all of the carbonized coating on these parts, exposing the untreated metal underneath. After this the shaft is heated in a fire and hardened, and as the parts *D*, *E*, and *F* are the only parts left carbonized, they will be the only parts hardened, leaving the rest soft.

TABLES.

TABLE I.

Circumferences and Areas of Circles.

Diameter.	Circumference.	Area.	Diameter.	Circumference.	Area.
1/4	.7854	.0490	3	9.4248	7.0686
5/16	.9817	.0767	1/8	9.8175	7.6699
3/8	1.1781	.1104	1/4	10.210	8.2958
7/16	1.3744	.1503	3/8	10.603	8.9462
1/2	1.5708	.1963	1/2	10.996	9.6211
9/16	1.7671	.2485	5/8	11.388	10.321
5/8	1.9635	.3068	3/4	11.781	11.045
11/16	2.1598	.3712	7/8	12.174	11.793
3/4	2.3562	.4417	4	12.566	12.566
13/16	2.5525	.5184	1/8	12.959	13.364
7/8	2.7489	.6013	1/4	13.352	14.186
15/16	2.9452	.6902	3/8	13.744	15.033
1	3.1416	.7854	1/2	14.137	15.904
1/16	3.3379	.8866	5/8	14.530	16.800
1/8	3.5343	.9940	3/4	14.923	17.728
3/16	3.7306	1.1075	7/8	15.315	18.665
1/4	3.9270	1.2272	5	15.708	19.635
5/16	4.1233	1.3530	1/8	16.101	20.629
3/8	4.3197	1.4849	1/4	16.493	21.648
7/16	4.5160	1.6230	3/8	16.886	22.691
1/2	4.7124	1.7671	1/2	17.279	23.758
9/16	4.9087	1.9175	5/8	17.671	24.850
5/8	5.1051	2.0739	3/4	18.064	25.967
11/16	5.3014	2.2365	7/8	18.457	27.109
3/4	5.4978	2.4053	6	18.850	28.274
13/16	5.6941	2.5802	1/8	19.242	29.465
7/8	5.8905	2.7612	1/4	19.635	30.680
15/16	6.0868	2.9483	3/8	20.028	31.919
2	6.2832	3.1416	1/2	20.420	33.183
1/16	6.4795	3.3410	5/8	20.813	34.472
1/8	6.6759	3.5466	3/4	21.206	35.785
3/16	6.8722	3.7583	7/8	21.598	37.122
1/4	7.0686	3.9761	7	21.991	38.485
5/16	7.2649	4.2000	1/8	22.384	39.871
3/8	7.4613	4.4301	1/4	22.776	41.282
7/16	7.6576	4.6664	3/8	23.169	42.718
1/2	7.8540	4.9087	1/2	23.562	44.179
9/16	8.0503	5.1572	5/8	23.955	45.664
5/8	8.2467	5.4119	3/4	24.347	47.173
11/16	8.4430	5.6727	7/8	24.740	48.707
3/4	8.6394	5.9396	8	25.133	50.265
13/16	8.8357	6.2126	1/8	25.525	51.849
7/8	9.0321	6.4918	1/4	25.918	53.456
15/16	9.2284	6.7771	3/8	26.311	55.088

TABLE I—(*Continued*).

CIRCUMFERENCES AND AREAS OF CIRCLES.

Diameter.	Circumference.	Area.	Diameter.	Circumference.	Area.
8½	26.704	56.745	16¼	51.051	207.39
⅝	27.096	58.426	½	51.836	213.82
¾	27.489	60.132	¾	52.622	220.35
⅞	27.882	61.862	17	53.407	226.98
9	28.274	63.617	¼	54.192	233.71
⅛	28.667	65.397	½	54.978	240.53
¼	29.060	67.201	¾	55.763	247.45
⅜	29.452	69.029	18	56.549	254.47
½	29.845	70.882	¼	57.334	261.59
⅝	30.238	72.760	½	58.119	268.80
¾	30.631	74.662	¾	58.905	276.12
⅞	31.023	76.589	19	59.690	283.53
10	31.416	78.540	¼	60.476	291.04
⅛	31.809	80.516	½	61.261	298.65
¼	32.201	82.516	¾	62.046	306.35
⅜	32.594	84.541	20	62.832	314.16
½	32.987	86.590	¼	63.617	322.06
⅝	33.379	88.664	½	64.403	330.06
¾	33.772	90.763	¾	65.188	338.16
⅞	34.165	92.886	21	65.973	346.36
11	34.558	95.033	¼	66.759	354.66
⅛	34.950	97.205	½	67.544	363.05
¼	35.343	99.402	¾	68.330	371.54
⅜	35.736	101.62	22	69.115	380.13
½	36.128	103.87	¼	69.900	388.82
⅝	36.521	106.14	½	70.686	397.61
¾	36.914	108.43	¾	71.471	406.49
⅞	37.306	110.75	23	72.257	415.48
12	37.699	113.10	¼	73.042	424.56
¼	38.485	117.86	½	73.827	433.74
½	39.270	122.72	¾	74.613	443.01
¾	40.055	127.68	24	75.398	452.39
13	40.841	132.73	¼	76.184	461.86
¼	41.626	137.89	½	76.969	471.44
½	42.412	143.14	¾	77.754	481.11
¾	43.197	148.49	25	78.540	490.87
14	43.982	153.94	¼	79.325	500.74
¼	44.768	159.48	½	80.111	510.71
½	45.553	165.13	¾	80.896	520.77
¾	46.338	170.87	26	81.681	530.93
15	47.124	176.71	¼	82.467	541.19
¼	47.909	182.65	½	83.252	551.55
½	48.695	188.69	¾	84.038	562.00
¾	49.480	194.83	27	84.823	572.56
16	50.265	201.06	¼	85.608	583.21

TABLE I—(*Continued*).

Circumferences and Areas of Circles.

Diameter.	Circumference.	Area.	Diameter.	Circumference.	Area.
27½	86.394	593.96	38¾	121.737	1179.3
¾	87.179	604.81	39	122.522	1194.6
28	87.965	615.75	¼	123.308	1210.0
¼	88.750	626.80	½	124.093	1225.4
½	89.535	637.94	¾	124.878	1241.0
¾	90.321	649.18	40	125.664	1256.6
29	91.106	660.52	¼	126.449	1272.4
¼	91.892	671.96	½	127.235	1288.2
½	92.677	683.49	¾	128.020	1304.2
¾	93.462	695.13	41	128.805	1320.3
30	94.248	706.86	¼	129.591	1336.4
¼	95.033	718.69	½	130.376	1352.7
½	95.819	730.62	¾	131.161	1369.0
¾	96.604	742.64	42	131.947	1385.4
31	97.389	754.77	¼	132.732	1402.0
¼	98.175	766.99	½	133.518	1418.6
½	98.960	779.31	¾	134.303	1435.4
¾	99.746	791.73	43	135.088	1452.2
32	100.531	804.25	¼	135.874	1469.1
¼	101.316	816.86	½	136.659	1486.2
½	102.102	829.58	¾	137.445	1503.3
¾	102.887	842.39	44	138.230	1520.5
33	103.673	855.30	¼	139.015	1537.9
¼	104.458	868.31	½	139.801	1555.3
½	105.243	881.41	¾	140.586	1572.8
¾	106.029	894.62	45	141.372	1590.4
34	106.814	907.92	¼	142.157	1608.2
¼	107.600	921.32	½	142.942	1626.0
½	108.385	934.82	¾	143.728	1643.9
¾	109.170	948.42	46	144.513	1661.9
35	109.956	962.11	¼	145.299	1680.0
¼	110.741	975.91	½	146.084	1698.2
½	111.527	989.80	¾	146.869	1716.5
¾	112.312	1003.8	47	147.655	1734.9
36	113.097	1017.9	¼	148.440	1753.5
¼	113.883	1032.1	½	149.226	1772.1
½	114.668	1046.3	¾	150.011	1790.8
¾	115.454	1060.7	48	150.796	1809.6
37	116.239	1075.2	¼	151.582	1828.5
¼	117.024	1089.8	½	152.367	1847.5
½	117.810	1104.5	¾	153.153	1866.5
¾	118.596	1119.2	49	153.938	1885.7
38	119.381	1134.1	¼	154.723	1905.0
¼	120.166	1149.1	½	155.509	1924.4
½	120.951	1164.2	¾	156.294	1943.9

TABLE I—(*Continued*).
Circumferences and Areas of Circles.

Diameter.	Circumference.	Area.	Diameter.	Circumference.	Area.
50	157.080	1963.5	62½	196.350	3068.0
¼	157.865	1983.2	63	197.920	3117.2
½	158.650	2003.0	½	199.491	3166.9
¾	159.436	2022.8	64	201.062	3217.0
51	160.221	2042.8	½	202.633	3267.5
¼	161.007	2062.9	65	204.204	3318.3
½	161.792	2083.1	½	205.774	3369.6
¾	162.577	2103.3	66	207.345	3421.2
52	163.363	2123.7	½	208.916	3473.2
¼	164.148	2144.2	67	210.487	3525.7
½	164.934	2164.8	½	212.058	3578.5
¾	165.719	2185.4	68	213.628	3631.7
53	166.504	2206.2	½	215.199	3685.3
¼	167.290	2227.0	69	216.770	3739.3
½	168.075	2248.0	½	218.341	3793.7
¾	168.861	2269.1	70	219.911	3848.5
54	169.646	2290.2	½	221.482	3903.6
¼	170.431	2311.5	71	223.053	3959.2
½	171.217	2332.8	½	224.624	4015.2
¾	172.002	2354.3	72	226.195	4071.5
55	172.788	2375.8	½	227.765	4128.2
¼	173.573	2397.5	73	229.336	4185.4
½	174.358	2419.2	½	230.907	4242.9
¾	175.144	2441.1	74	232.478	4300.8
56	175.929	2463.0	½	234.049	4359.2
¼	176.715	2485.0	75	235.619	4417.9
½	177.500	2507.2	½	237.190	4477.0
¾	178.285	2529.4	76	238.761	4536.5
57	179.071	2551.8	½	240.332	4596.3
¼	179.856	2574.2	77	241.903	4656.6
½	180.642	2596.7	½	243.473	4717.3
¾	181.427	2619.4	78	245.044	4778.4
58	182.212	2642.1	½	246.615	4839.8
¼	182.998	2664.9	79	248.186	4901.7
½	183.783	2687.8	½	249.757	4963.9
¾	184.569	2710.9	80	251.327	5026.5
59	185.354	2734.0	½	252.898	5089.6
¼	186.139	2757.2	81	254.469	5153.0
½	186.925	2780.5	½	256.040	5216.8
¾	187.710	2803.9	82	257.611	5281.0
60	188.496	2827.4	½	259.181	5345.6
½	190.066	2874.8	83	260.752	5410.6
61	191.637	2922.5	½	262.323	5476.0
½	193.208	2970.6	84	263.894	5541.8
62	194.779	3019.1	½	265.465	5607.9

TABLE I—(*Continued*).

Circumferences and Areas of Circles.

Diameter.	Circumference.	Area.	Diameter.	Circumference.	Area.
85	267.035	5674.5	93	292.168	6792.9
½	268.606	5741.5	½	293.739	6866.1
86	270.177	5808.8	94	295.310	6939.8
½	271.748	5876.5	½	296.881	7013.8
87	273.319	5944.7	95	298.451	7088.2
½	274.889	6013.2	½	300.022	7163.0
88	276.460	6082.1	96	301.593	7238.2
½	278.031	6151.4	½	303.164	7313.8
89	279.602	6221.1	97	304.734	7389.8
½	281.173	6291.2	½	306.305	7466.2
90	282.743	6361.7	98	307.876	7543.0
½	284.314	6432.6	½	309.447	7620.1
91	285.885	6503.9	99	311.018	7697.7
½	287.456	6575.5	½	312.588	7775.6
92	289.027	6647.6	100	314.159	7854.0
½	290.597	6720.1			

TABLE II.

TEMPERATURES TO WHICH HARDENED TOOLS SHOULD BE HEATED TO PROPERLY "DRAW THE TEMPER," TOGETHER WITH THE COLORS OF SCALE APPEARING ON A POLISHED-STEEL SURFACE AT THOSE TEMPERATURES, AND OTHER MEANS OF DETECTING PROPER HEATING.

Kind of Tool.	Temperature, Fahr.	Color of Scale.	Action of File.	Other Indications.
Scrapers for ordinary use.	200°			Water dries quickly.
Burnishers.		Very pale yellow.	Can hardly be made to catch.	Lard-oil smokes slightly.
Light turning and finishing tools.	430°			
Engraving-tools.			Can be made to catch with difficulty.	
Lathe-tools.				
Milling-cutters.	460°	Straw-yellow.		
Lathe- and planer-tools for heavy work.				
Taps.				
Dies for screw-cut'g.				
Reamers.				
Punches.				
Dies.				
Flat drills.				
Wood-working tools.	500°	Brown-yellow.		
Plane-irons.				
Wood-chisels.				
Wood-turning tools.				
Twist drills.				
Sledges.				
Bl'ksmiths' ham'rs.				
Cold-chisels for very light work.	530°	Light purple.	Scratches.	
Axes.	550°	Dark purple.		
Cold-chisels for ordinary use.		Blue, ting'd slightly with red.		
Stone-cutting chisels			Files with great difficulty.	
Carving-knives.				
Screw-drivers.				
Saws.				
Springs.	580°	Blue.	Files with difficulty.	Lard-oil burns or flashes.
	610°	Pale blue.		
	630°	Greenish blue.		

TABLE III.

Decimal Equivalents of Fractions of One Inch.

From Kent's Mechanical Engineer's Pocket-book.

1/64	.015625	33/64	.515625
1/32	.03125	17/32	.53125
3/64	.046875	35/64	.546875
1/16	.0625	9/16	.5625
5/64	.078125	37/64	.578125
3/32	.09375	19/32	.59375
7/64	.109375	39/64	.609375
1/8	.125	5/8	.625
9/64	.140625	41/64	.640625
5/32	.15625	21/32	.65625
11/64	.171875	43/64	.671875
3/16	.1875	11/16	.6875
13/64	.203125	45/64	.703125
7/32	.21875	23/32	.71875
15/64	.234375	47/64	.734375
1/4	.25	3/4	.75
17/64	.265625	49/64	.765625
9/32	.28125	25/32	.78125
19/64	.296875	51/64	.796875
5/16	.3125	13/16	.8125
21/64	.328125	53/64	.828125
11/32	.34375	27/32	.84375
23/64	.359375	55/64	.859375
3/8	.375	7/8	.875
25/64	.390625	57/64	.890625
13/32	.40625	29/32	.90625
27/64	.421875	59/64	.921875
7/16	.4375	15/16	.9375
29/64	.453125	61/64	.953125
15/32	.46875	31/32	.96875
31/64	.484375	63/64	.984375
1/2	.50	1	1.

TABLE IV.

Weights of Bar Steel per Lineal Foot.

The weight given in the table is for a bar of steel 1 foot long and of the dimensions named.

(From Jones & Laughlins.)

Size in Inches.	Rounds, Weight in Lbs.	Squares, Weight in Lbs.	Size in Inches.	Roun'ds, Weight in Lbs.	Squares, Weight in Lbs.	Width in Inches.	Thickness in Inches.											
							1/16	1/8	3/16	1/4	5/16	3/8	7/16	1/2	5/8	3/4	7/8	1
3/16	.094	.120	2 1/8	12.06	15.36	1	.21	.43	.638	.850	1.06	1.28	1.49	1.70	2.12	2.55	2.98	
1/4	.167	.213	2 1/4	13.52	17.22	1 1/8	.24	.48	.720	.955	1.20	1.43	1.68	1.92	2.39	2.87	3.35	3.88
5/16	.261	.332	2 3/8	15.07	19.19	1 1/4	.27	.53	.797	1.06	1.33	1.59	1.86	2.12	2.65	3.19	3.72	4.25
3/8	.375	.478	2 1/2	16.70	21.26	1 3/8	.30	.59	.875	1.17	1.46	1.76	2.05	2.34	2.92	3.51	4.09	4.68
7/16	.511	.651	2 5/8	18.41	23.44	1 1/2	.32	.64	.957	1.28	1.59	1.92	2.23	2.55	3.19	3.83	4.47	5.10
1/2	.668	.851	2 3/4	20.21	25.73	1 5/8	.35	.69	1.04	1.38	1.73	2.08	2.42	2.77	3.46	4.15	4.84	5.53
9/16	.845	1.076	3	24.05	30.62	1 3/4	.38	.75	1.11	1.49	1.86	2.23	2.60	2.98	3.72	4.47	5.20	5.95
5/8	1.044	1.329	3 1/4	28.23	35.94	2	.43	.85	1.28	1.70	2.12	2.55	2.98	3.40	4.25	5.10	5.95	6.80
3/4	1.503	1.914	3 1/2	32.7[illegible]	41.68	2 1/4	.48	.96	1.44	1.91	2.39	2.87	3.35	3.83	4.78	5.75	6.69	7.65
7/8	2.046	2.605	3 3/4	37.57	47.84	2 1/2	.53	1.06	1.59	2.12	2.65	3.19	3.72	4.25	5.31	6.38	7.44	8.50
1	2.672	3.402	4	42.77	54.45	2 3/4	.59	1.17	1.75	2.34	2.92	3.51	4.09	4.67	5.84	7.02	8.18	9.35
1 1/8	3.382	4.306	4 1/2	54.83	69.81	3	.64	1.28	1.91	2.55	3.19	3.83	4.46	5.10	6.38	7.65	8.93	10.20
1 1/4	4.175	5.316	5	66.82	85.08	3 1/4	.69	1.38	2.07	2.76	3.45	4.15	4.83	5.53	6.91	8.29	9.67	11.05
1 3/8	5.052	6.432	5 1/2	80.85	102.94	3 1/2	.75	1.49	2.23	2.98	3.72	4.47	5.20	5.95	7.44	8.93	10.41	11.90
1 1/2	6.012	7.655	6	96.22	122.51	3 3/4	.80	1.60	2.39	3.19	3.99	4.78	5.58	6.38	7.97	9.57	11.16	12.75
1 5/8	7.056	8.984	6 1/2	112.92	143.78	4	.85	1.70	2.55	3.40	4.25	5.10	5.95	6.80	8.50	10.20	11.90	13.60
1 3/4	8.183	10.419	7	130.97	166.75	4 1/2	.96	1.92	2.87	3.83	4.78	5.74	6.70	7.65	9.57	11.48	13.39	15.30
1 7/8	9.394	11.961	7 1/2	150.34	191.42	5	1.07	2.13	3.19	4.25	5.31	6.38	7.44	8.50	10.63	12.75	14.87	17.00
2	10.69	13.61	8	171.04	217.78	5 1/2	1.17	2.34	3.51	4.67	5.84	7.02	8.18	9.35	11.69	14.03	16.36	18.70
						6	1.28	2.55	3.83	5.10	6.38	7.65	8.93	10.20	12.75	15.30	17.85	20.40
						7	1.49	2.98	4.46	5.95	7.44	8.93	10.41	11.90	14.87	17.85	20.83	23.80
						8	1.70	3.40	5.10	6.80	8.50	10.20	11.90	13.60	17.00	20.40	23.80	27.20

COURSE OF EXERCISES IN FORGE WORK.

What is suggested as a standard course of exercises is given below.

A short talk should first be given covering the calculation of stock for simple bent work, rings, links, eyes, etc.

The starting of the fire and fitting of tongs is then explained.

Exercise 1. Stock $\frac{1}{2}'' \times \frac{1}{2}'' \times 6''$ is drawn out to $\frac{1}{4}''$ round, and this round stock is used to make the two following pieces of work.

Exercise 2. Fig. 273. Eye Bend.

Exercise 3. Fig. 274. Double Eye Bend.

Exercise 4. Fig. 276. Twisted Gate Hook.

Exercise 5. Fig. 275. Square Point and Eye Bend.

Exercise 6. Fig. 312. Twisted Scriber. Before giving this exercise a short talk should be given on the effect of high heats on tool steel. The scriber should be forged from an old file in order to give practice in the drawing out of tool steel. The scriber is tempered later in the course.

Exercise 7. Fig. 277. Weldless Ring.

Exercise 8. Fig. 290. Chain Hook.

Exercise 9. Fig. 286. Bracket with Forged Corner.

Exercise 10. Practice Weld. This should be a sort of a faggot weld made by doubling over the end of a piece of scrap, the object being simply to determine the welding heat.

Exercise 11. Fig. 292. Chain of Three Links.

Exercise 12. Fig. 283. Flat Lap Weld.

Exercise 13. Fig. 284. Angle Weld.

Exercise 14. Fig. 282. Welded Ring. This and the hook made in Ex. 9 should each be joined to the chain by extra links, making a chain of five links with the hook on one end and the ring on the other.

Exercise 15. Fig. 308. Welded Rings shrunk together.

Exercise 16. Fig. 293. Planer Bolt, make welded head.

Exercise 17. Fig. 287. Hexagonal Head Bolt, upset head.

Exercise 18. Fig. 289. Ladle.

Exercise 19. Fig. 302. Taper Machine Key.

Exercise 20. Fig. 303. Lever Arm.

Exercise 21. Figs. 305, 306, or 307. Tongs.

Exercise 22. Hardening Tool Steel. The student should be given an old file or piece of scrap tool steel to determine the proper hardening heat. This is done by drawing out the steel to about $\frac{1}{4}''$ square and hardening the end, which is then snapped off and the condition of the steel determined from the fracture. This should be repeated until the hardening heat can be hit upon every time.

Exercise 23. Fig. 313. Cold Chisel.

Exercise 24. Fig. 315. Center Punch.
Exercise 25. Fig. 332. Cape Chisel.
Exercise 26. Figs. 317 or 318. Thread Tool.
Exercise 27. Fig. 319. Round Nose Tool.
Exercise 28. Fig. 316. Side Tool.
Exercise 29. Fig. 322. Boring Tool.
Exercise 30. Fig. 320. Diamond Point.
Exercise 31. Figs. 328, 329, 330, 331, or 334. Hot Chisel, Cold Chisel, Set Hammer, Flatter or Pattern-maker's Hammer.
Exercise 32. Fig. 335. Spring.
Exercise 33. Fig. 310. Brazed Ring.

Many students will be able to cover much more ground than outlined above, and for such cases additional drawings are given. These additional exercises may be interpolated where the instructor sees fit.

Additional drawings are also given in order that the course may be varied somewhat from term to term.

No more than three pieces of stock should ever be allowed for any one exercise, and as a general rule the student should do the work with one. When more than one piece is used the work should be graded down accordingly.

Talks should be given on Brazing, Case Hardening, Metallurgy of Bessemer, Open Hearth, and Crucible Steels and Wrought Iron.

Considerable work should also be done in making sketches and stock calculations for large machine forgings, the sketches to show the different steps in the forging process.

When a steam or power hammer is available old hammers, tools, etc., may be drawn out into bar stock for center punches, small chisels, etc.

The tongs shown in the drawings may be made to good advantage under a steam or power hammer.

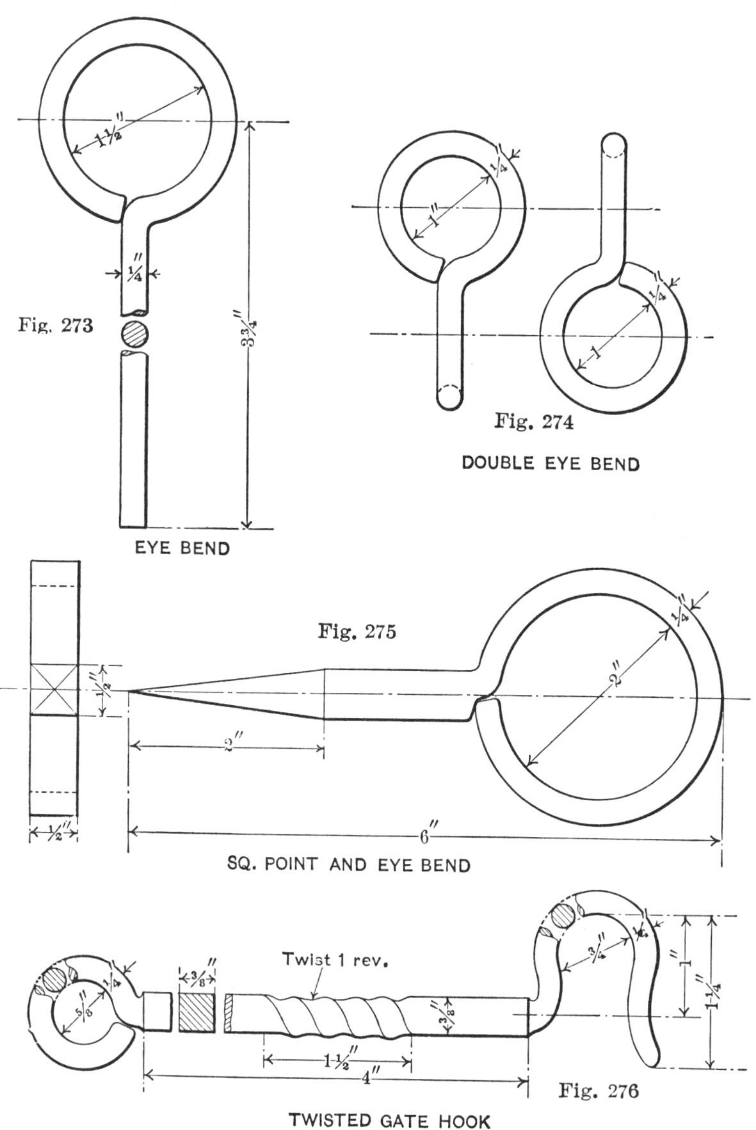

Fig. 273

Fig. 274

Fig. 275

Fig. 276

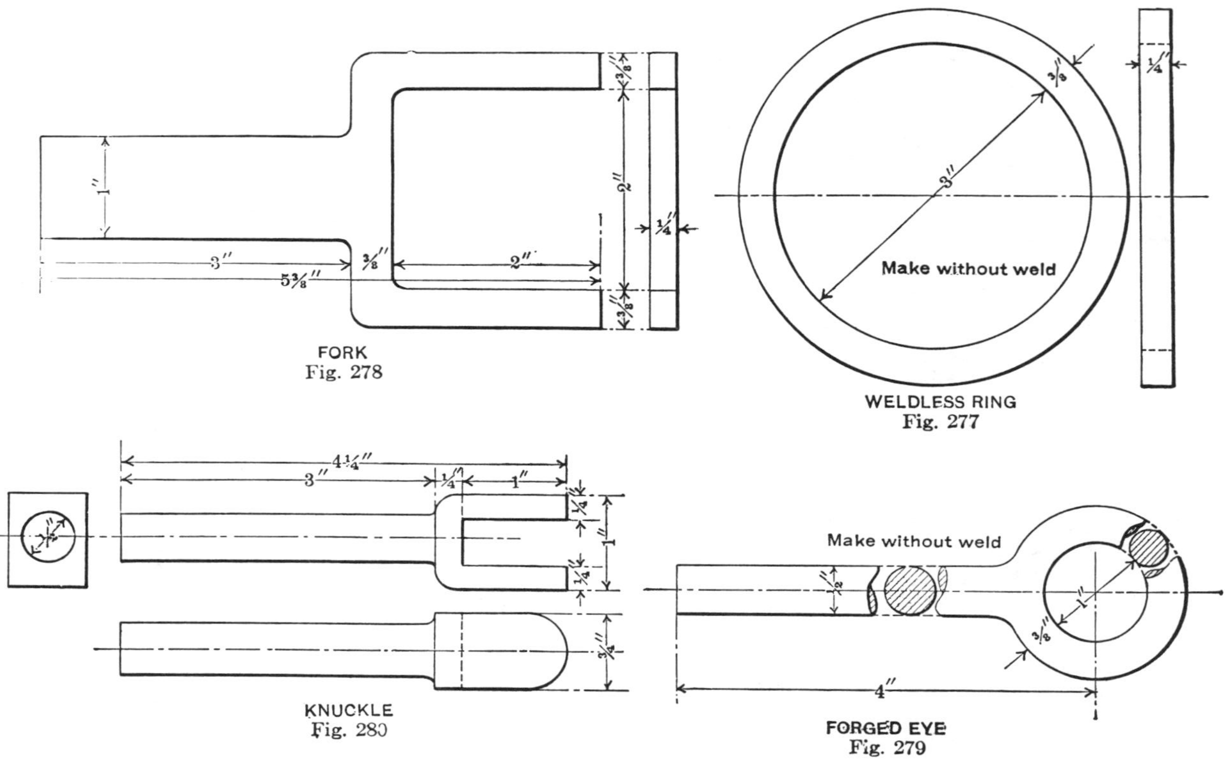

FORK
Fig. 278

WELDLESS RING
Fig. 277

KNUCKLE
Fig. 280

FORGED EYE
Fig. 279

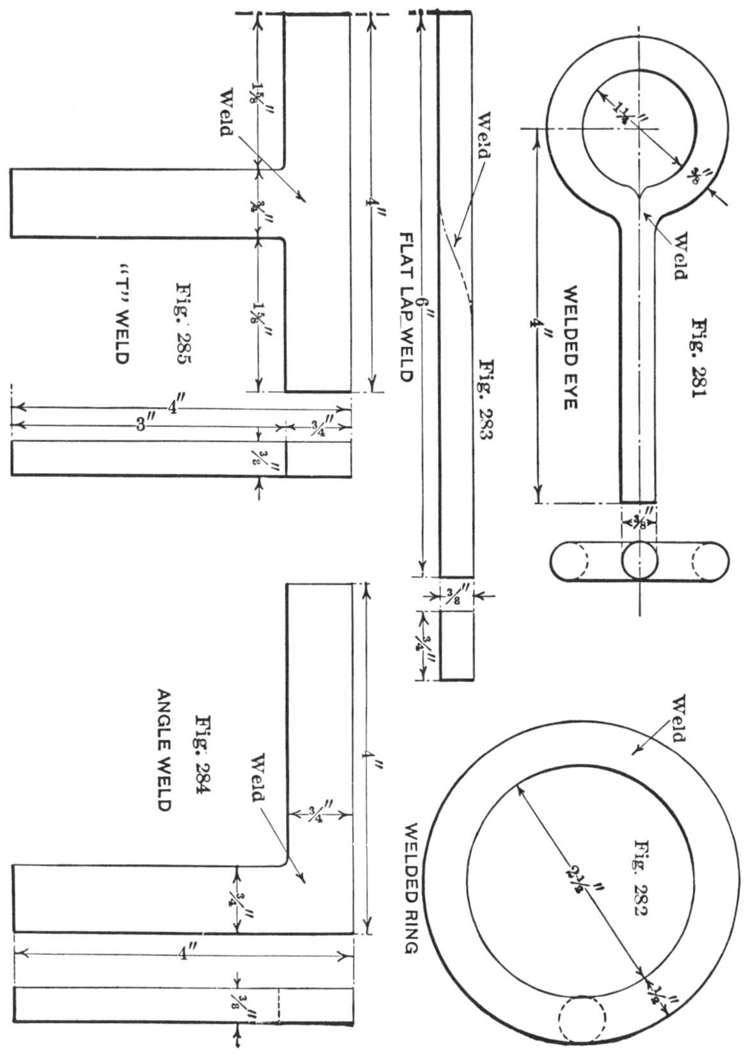

Fig. 281 WELDED EYE

Fig. 282 WELDED RING

Fig. 283 FLAT LAP WELD

Fig. 284 ANGLE WELD

Fig. 285 "T" WELD

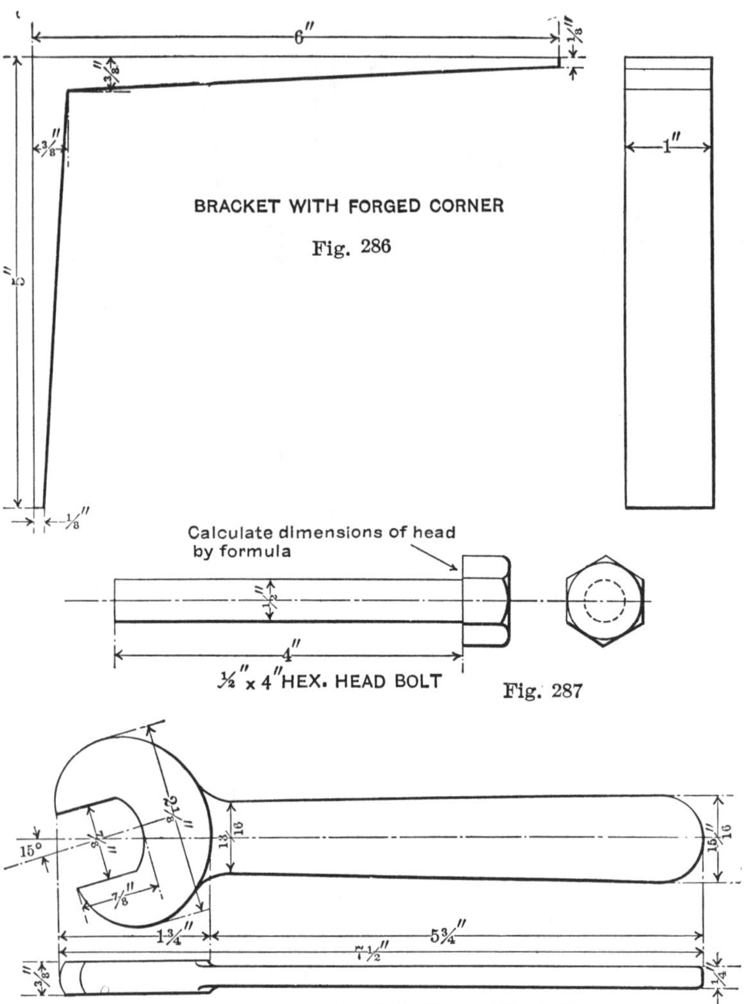

BRACKET WITH FORGED CORNER

Fig. 286

½" x 4" HEX. HEAD BOLT

Fig. 287

OPEN WRENCH

Fig. 288

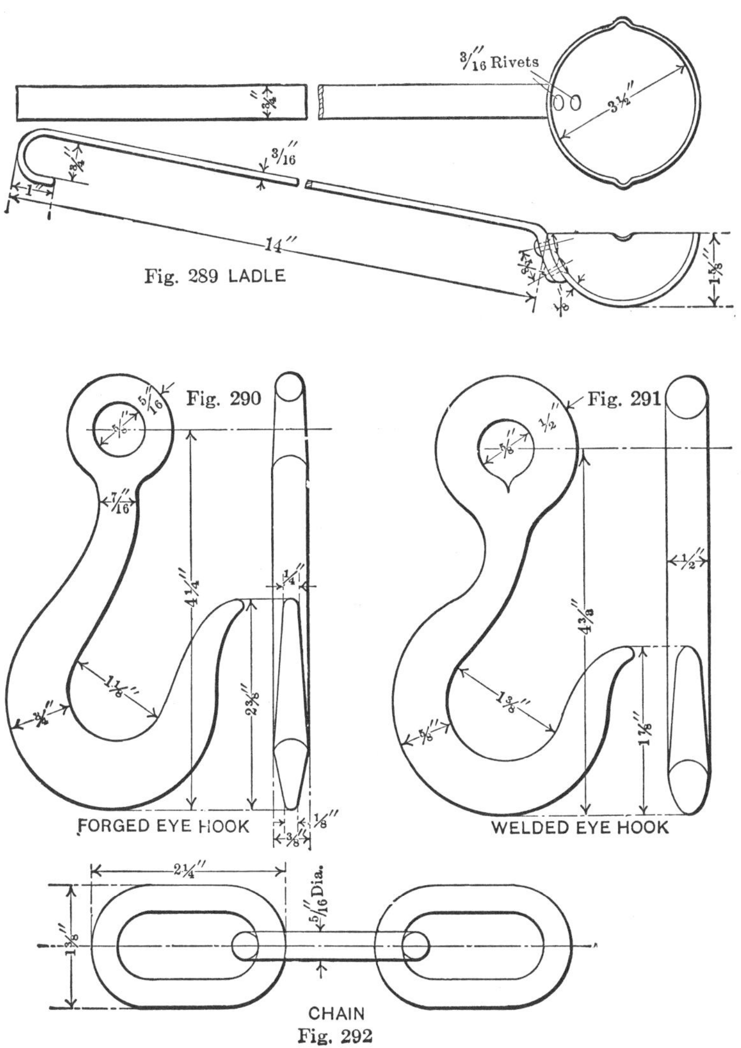

Fig. 289 LADLE

CHAIN
Fig. 292

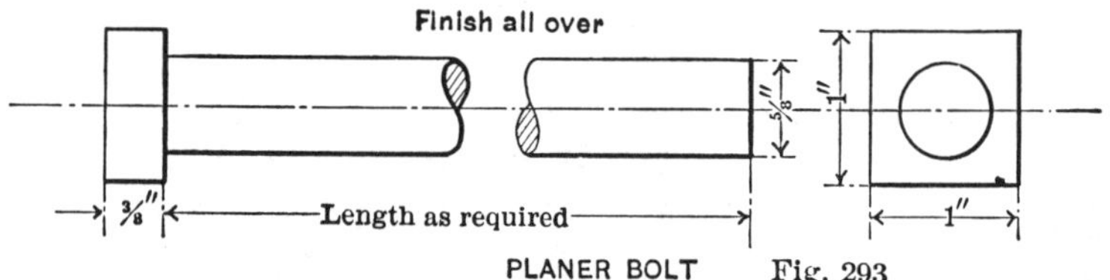

PLANER BOLT Fig. 293

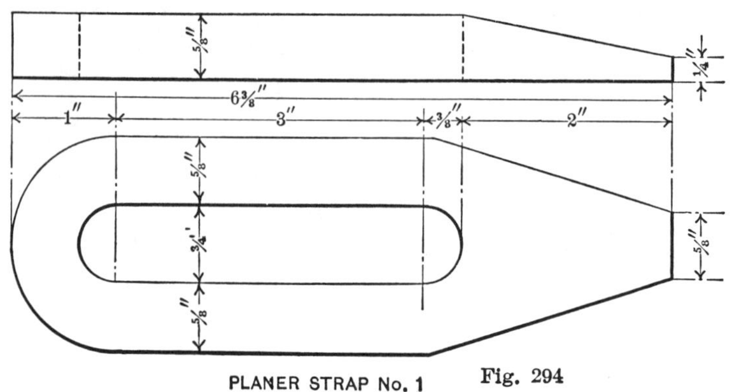

PLANER STRAP No. 1 Fig. 294

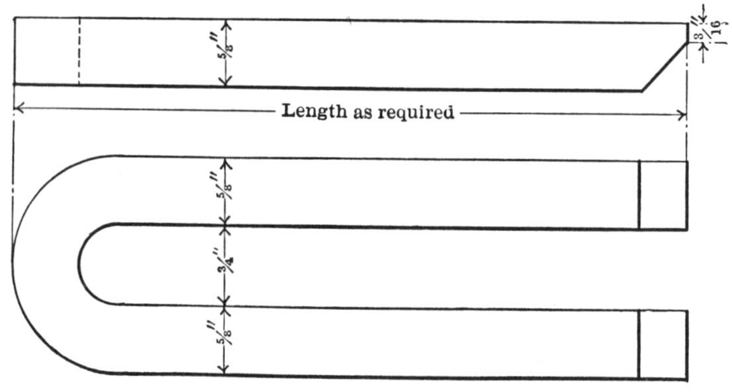

PLANER STRAP No. 2

Fig. 295

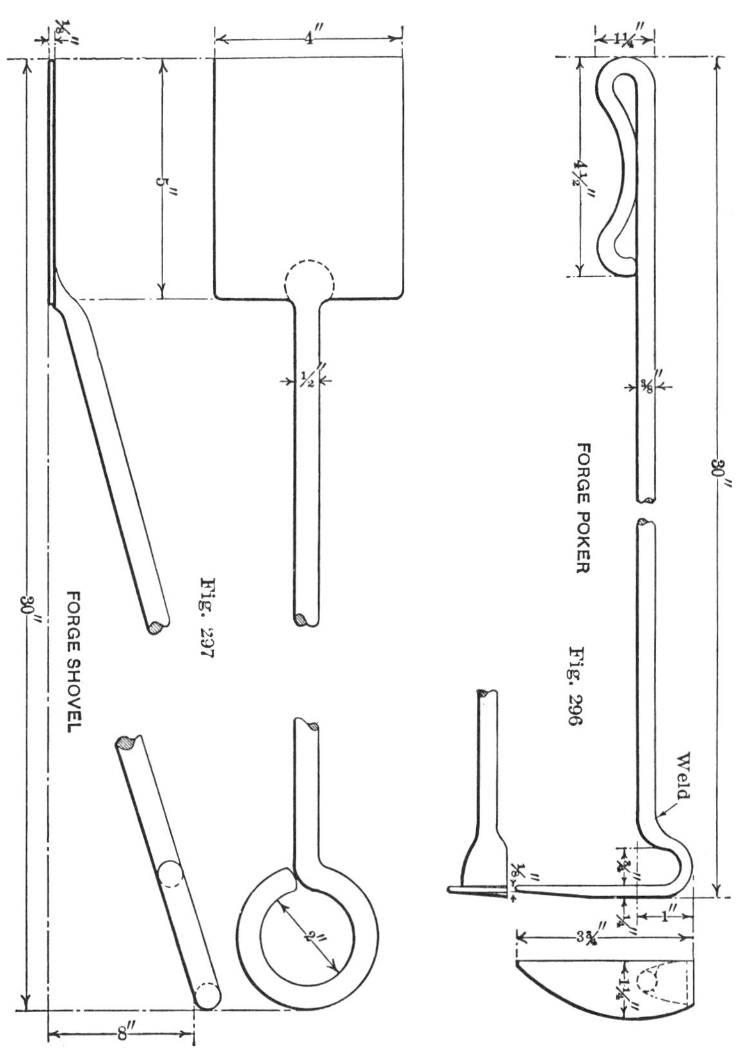

Fig. 297

Fig. 296

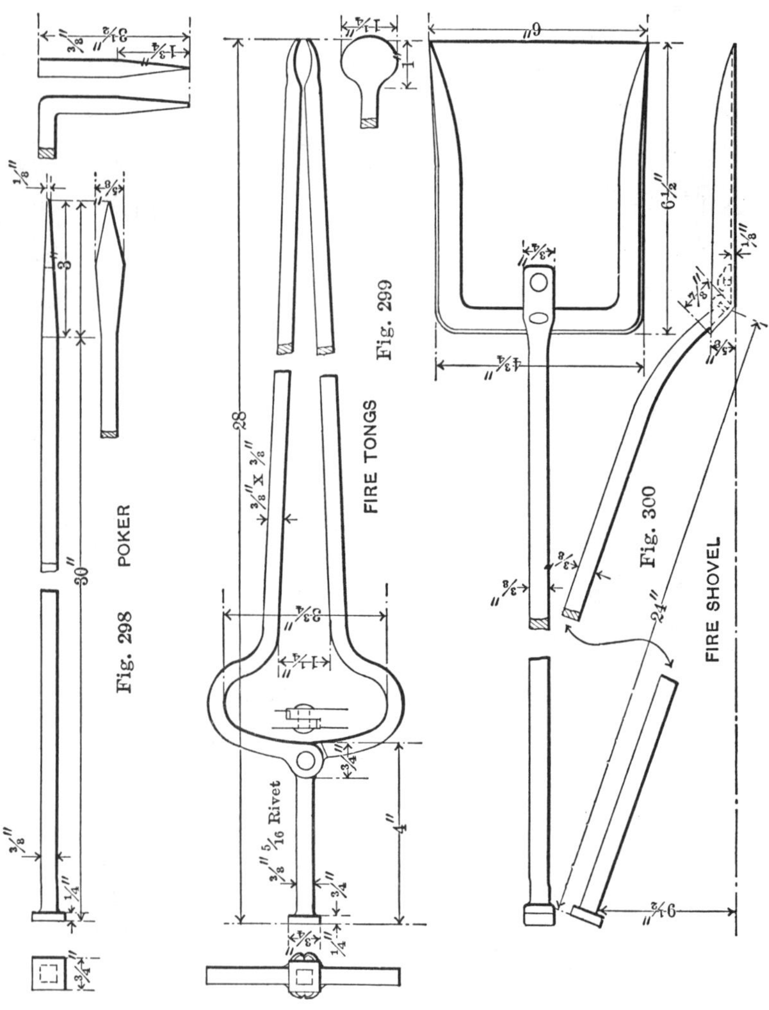

Fig. 298 POKER

Fig. 299 FIRE TONGS

Fig. 300 FIRE SHOVEL

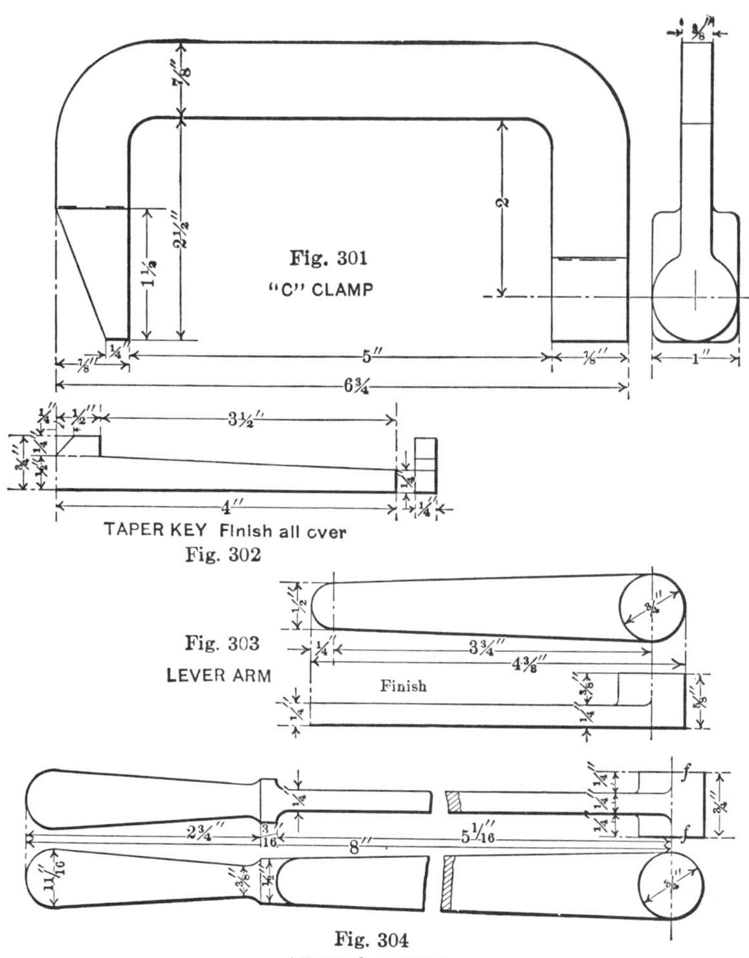

Fig. 301
"C" CLAMP

TAPER KEY Finish all over
Fig. 302

Fig. 303
LEVER ARM

Fig. 304
LEVER & HANDLE

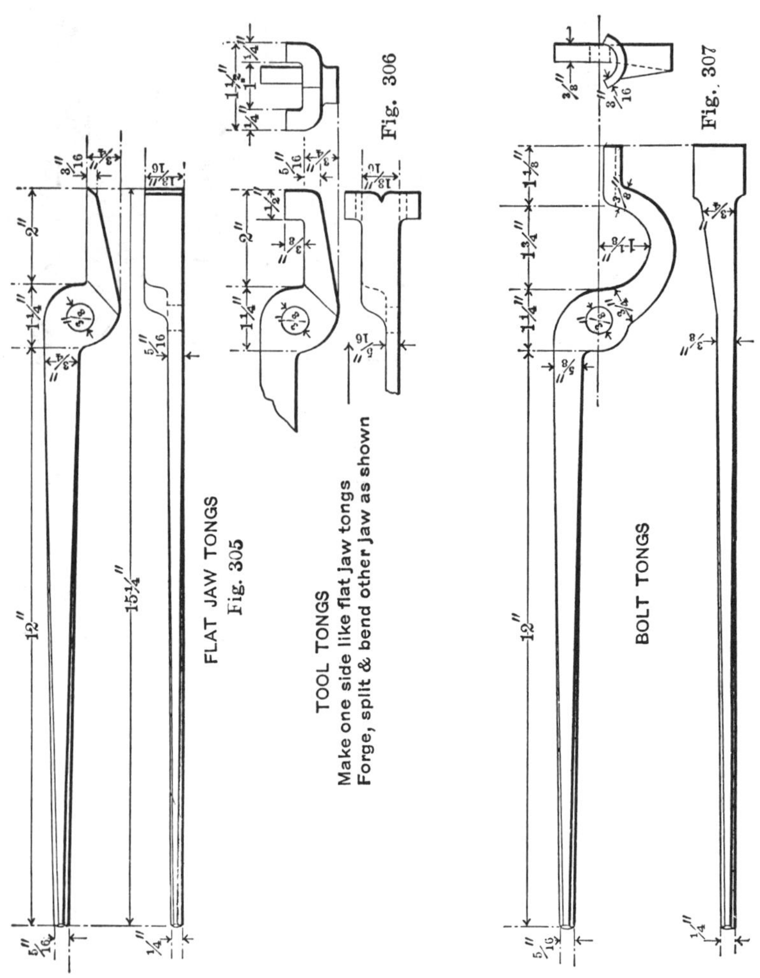

FLAT JAW TONGS
Fig. 305

TOOL TONGS
Make one side like flat jaw tongs
Forge, split & bend other jaw as shown
Fig. 306

BOLT TONGS
Fig. 307

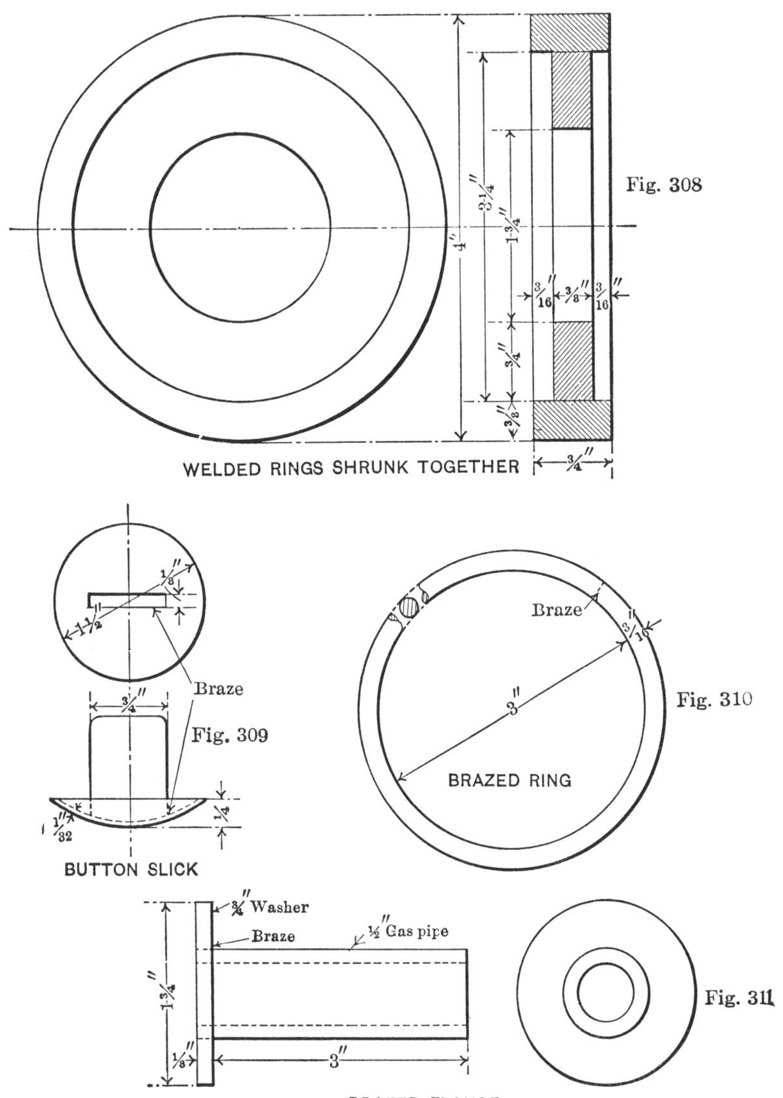
Fig. 308
WELDED RINGS SHRUNK TOGETHER
Braze
Fig. 309
BUTTON SLICK
Braze
Fig. 310
BRAZED RING
¾" Washer
Braze
½" Gas pipe
Fig. 311
BRAZED FLANGE

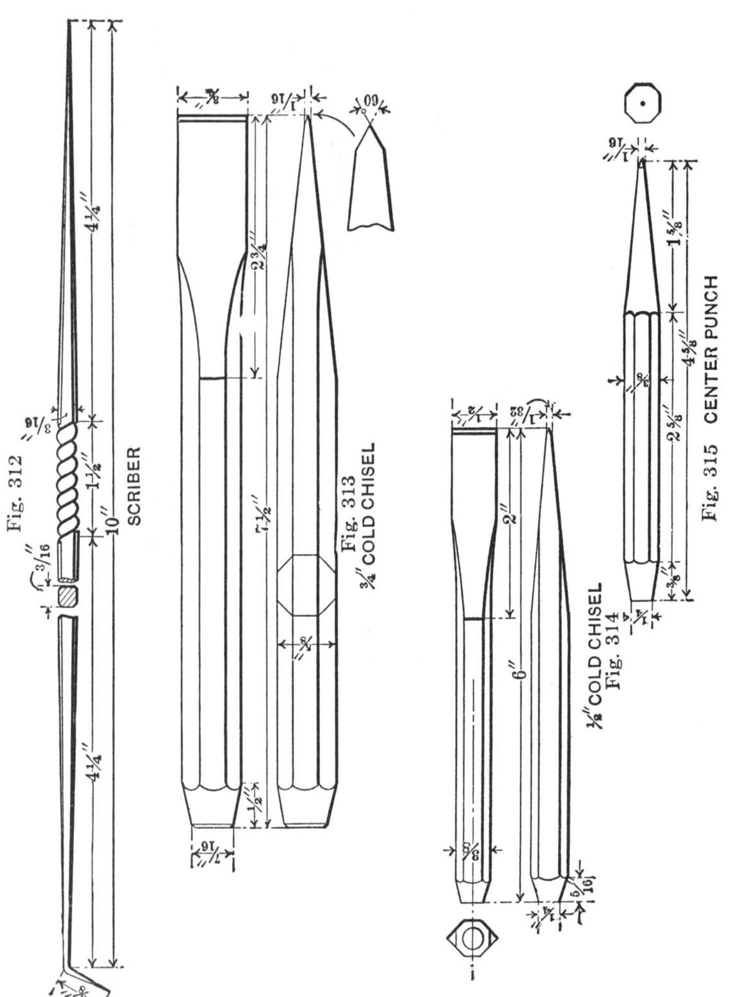

Fig. 312 SCRIBER

Fig. 313 3/4" COLD CHISEL

1/2" COLD CHISEL Fig. 314

Fig. 315 CENTER PUNCH

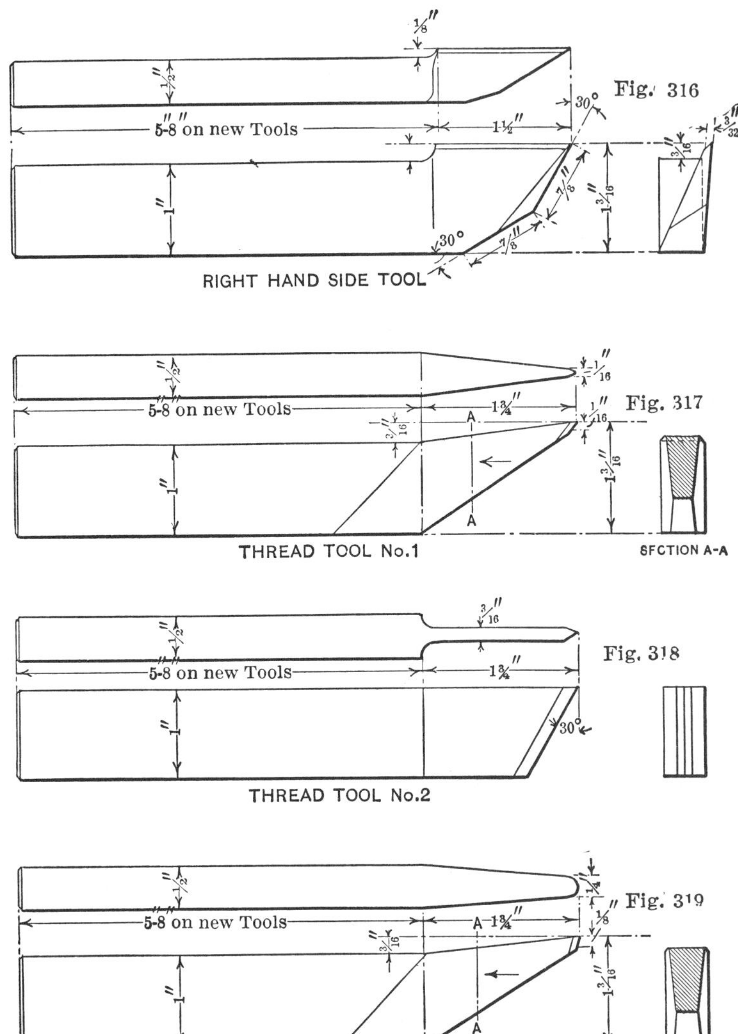

RIGHT HAND SIDE TOOL

Fig. 316

THREAD TOOL No.1

Fig. 317

THREAD TOOL No.2

Fig. 318

ROUND NOSE TOOL

Fig. 319

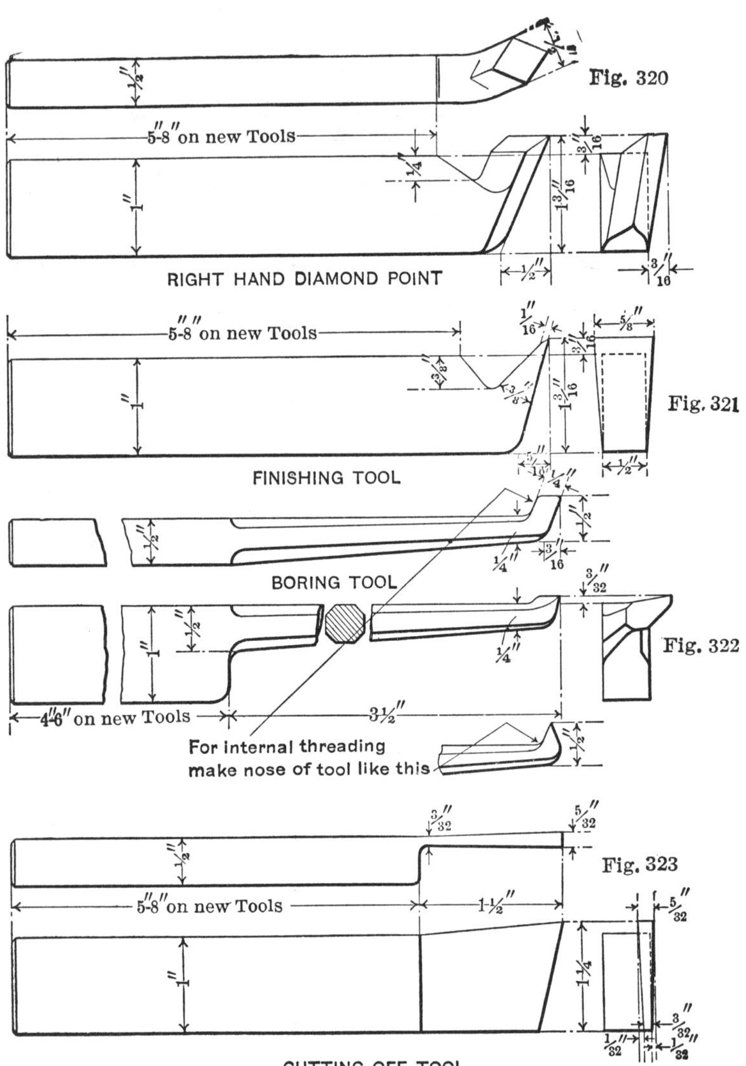
Fig. 320
5″-8″ on new Tools
RIGHT HAND DIAMOND POINT
5″-8″ on new Tools
Fig. 321
FINISHING TOOL
BORING TOOL
Fig. 322
4″-6″ on new Tools
3½″
For internal threading
make nose of tool like this
Fig. 323
5″-8″ on new Tools
1½″
CUTTING OFF TOOL

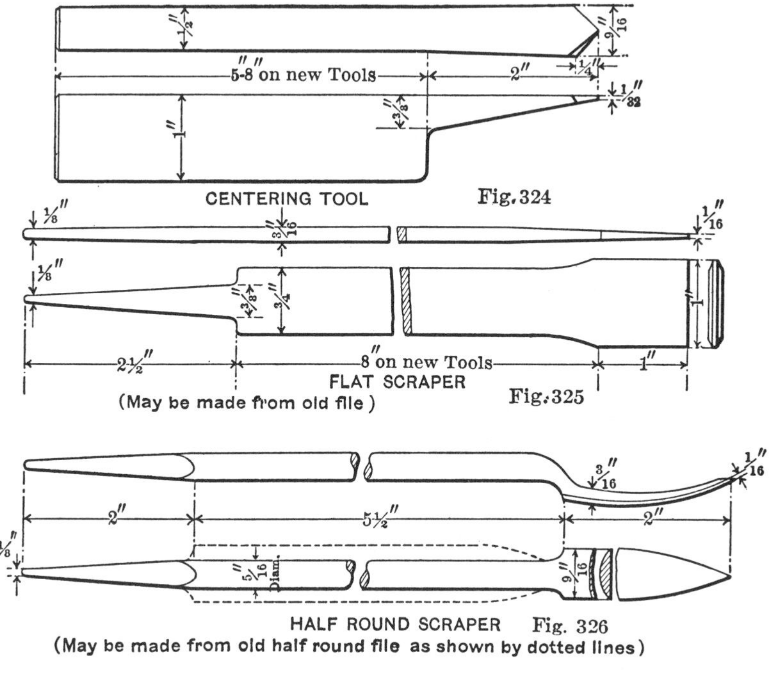

CENTERING TOOL Fig. 324

FLAT SCRAPER (May be made from old file) Fig. 325

HALF ROUND SCRAPER Fig. 326 (May be made from old half round file as shown by dotted lines)

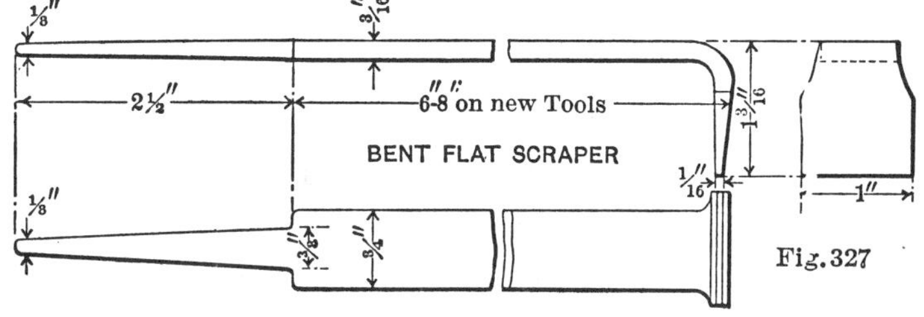

Fig. 327

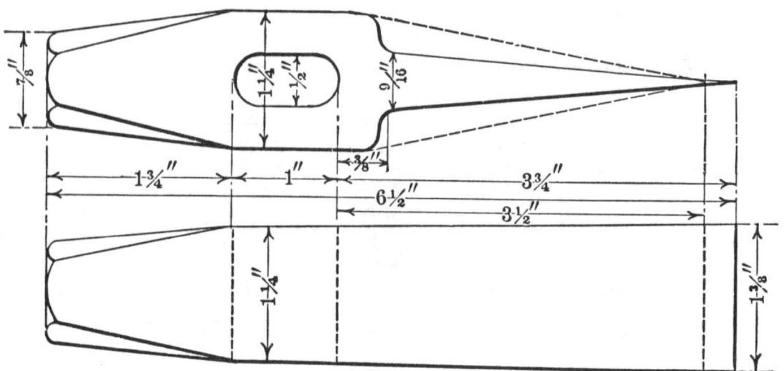

Fig. 328 BLACKSMITH'S HOT CHISEL

Fig. 329 BLACKSMITH'S COLD CHISEL
Make same way but shape end as shown by dotted line.

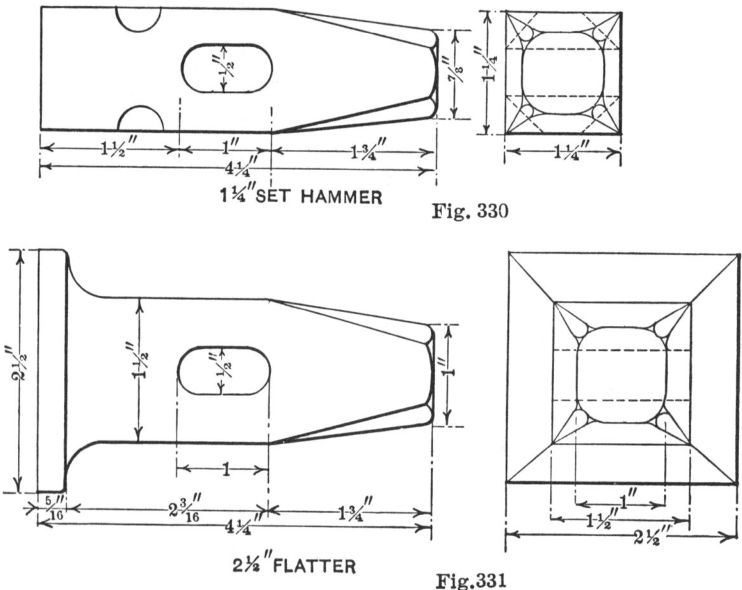

1¼″ SET HAMMER

Fig. 330

2½″ FLATTER

Fig. 331

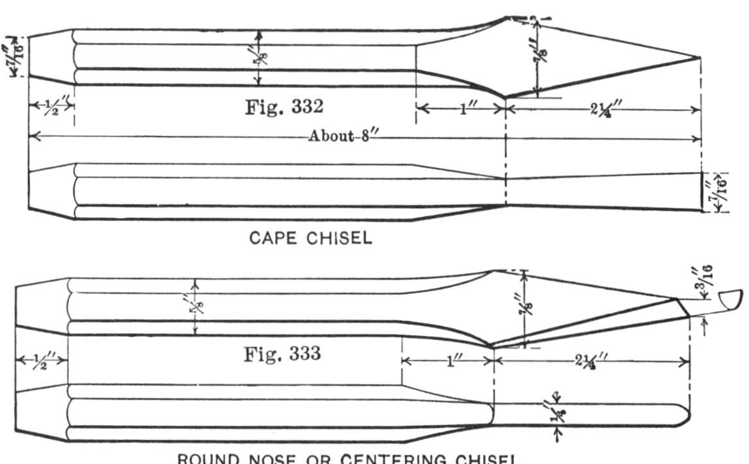

CAPE CHISEL

ROUND NOSE OR CENTERING CHISEL

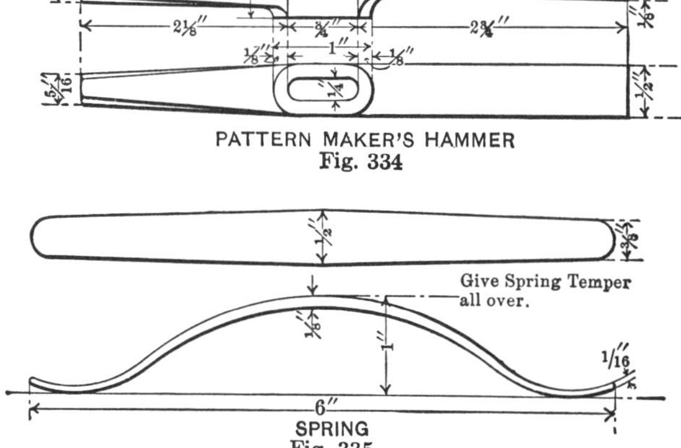

PATTERN MAKER'S HAMMER
Fig. 334

SPRING
Fig. 335

INDEX.